A Circular Economy for Manufacturing Waste Management

Pathway Towards Sustainability

RIVER PUBLISHERS SERIES IN ENERGY SUSTAINABILITY AND EFFICIENCY

Series Editor:

PEDRAM ASEF
Lecturer (Asst. Prof.) in Automotive Engineering
University of Hertfordshire
UK

The "River Publishers Series in Sustainability and Efficiency" is a series of comprehensive academic and professional books which focus on theory and applications in sustainable and efficient energy solutions. The books serve as a multi-disciplinary resource linking sustainable energy and society, fulfilling the rapidly growing worldwide interest in energy solutions. All fields of possible sustainable energy solutions and applications are addressed, not only from a technical point of view, but also from economic, social, political, and financial aspects. Books published in the series include research monographs, edited volumes, handbooks and textbooks. They provide professionals, researchers, educators, and advanced students in the field with an invaluable insight into the latest research and developments.

Topics covered in the series include, but are not limited to:

- Sustainable energy development and management;
- Alternate and renewable energies;
- Energy conservation;
- Energy efficiency;
- Carbon reduction;
- Environment.

For a list of other books in this series, visit www.riverpublishers.com

A Circular Economy for Manufacturing Waste Management

Pathway Towards Sustainability

Editors

Rakesh Kumar

Department of Management Studies,
Graphic Era Deemed to be University, Dehradun, India

Sachi Nandan Mohanty

School of Computer Science & Engineering (SCOPE),
VIT-AP University, Amaravati, Andhra Pradesh, India

Isaías Scalabrin Bianchi

Federal University of Santa Catarina,
Brazil

River Publishers

Routledge
Taylor & Francis Group
NEW YORK AND LONDON

Published 2026 by River Publishers
River Publishers
Broagervej 10, 9260 Gistrup, Denmark
www.riverpublishers.com

Distributed exclusively by Routledge
605 Third Avenue, New York, NY 10158, USA
4 Park Square, Milton Park, Abingdon, Oxon OX14 4RN

A Circular Economy for Manufacturing Waste Management / by Rakesh Kumar, Sachi Nandan Mohanty, Isaías Scalabrin Bianchi.

Routledge is an imprint of the Taylor & Francis Group, an informa business

ISBN 978-87-4380-689-9 (hardback)
ISBN 978-87-4380-692-9 (paperback)
ISBN 978-87-4380-691-2 (online)
ISBN 978-87-4380-690-5 (master ebook)

DOI: 10.1201/9788743806905

Contents

Preface

The book titled “Circular Economy for Manufacturing Waste Management: Pathway Towards Sustainability: explores the impact of circular economy ideas on waste management in manufacturing industries. It highlights the need for a shift from linear to circular models, reducing waste, recycling materials, and creating efficient goods. The book discusses the use of technologies like machine learning, IoT, and artificial intelligence in improving recycling and garbage monitoring. The book highlights the financial benefits of implementing circular economy principles, including reduced production costs and environmental compliance. The circular economy (CE) is an alternative to traditional take-make-dispose approaches, focusing on resource efficiency, waste reduction, and extending product use. This book explores smart technology applications in recycling, including human–machine interaction and predictive analysis. AI-based waste sorting, predictive maintenance in recycling centers, and future robotics are discussed to identify sustainability.Waste management is a global contest that requires innovative technologies to promote sustainability and a circular economy. Artificial intelligence (AI) offers effective solutions to address this complex issue, particularly within manufacturing companies. This book highlights the role of AI in waste management, including assembly, arrangement, reprocessing, and controlling. It demarcates the possible advantages and problems of using AI and highlights the enhanced quality of information, confidentiality, reduction of cost, and moral deliberations. AI holds a substantial role in effective waste management by solving issues like confidentiality concerns, data quality, and reduction of cost.The AI-driven waste minimization model in manufacturing is investigated by reviewing literature on their use and identified gaps in research. Intelligent systems are essential for sustainable waste management, as they offer a radical change toward a future based on ecologically robust resources. The book highlights sustainable supply chain management (SSCM) through smart technology, environmental, and social performance. Smart technologies like AI, IoT, blockchain, and big data analytics enhance SSCM by improving productivity, reducing waste, and increasing transparency. AI-powered predictive models support reduced inventory costs, avert overproduction, and estimate demand fluctuations accurately, resulting in a positive environmental impact. Blockchain technology ensures data integrity and traceability, increasing stakeholder accountability and confidence.

It examines corporate social responsibility (CSR) policies for circular economy in manufacturing industry, highlighting ethical, legal, and philanthropic principles through Carroll's Pyramid, stakeholder theory, triple bottom line, and shared value theory. The chapter explores managerial approaches such as sustainable supply chain management, eco-design innovation, and reverse logistics to recover waste value through product-as-a-service circular business models.

Challenges include intensive startup costs, regulatory difficulties, opposition from organizations and supply chain issues, and technological barriers that block total CSR and circular economy practice integration. The book provided insights for professionals, policymakers, and scholars.

Chapter 1 titled "Introduction to Circular Economy and its Applications in the Manufacturing Process," explores circular economic development and environmental balance. The circular economy (CE) represents an alternative to the traditional take-make-dispose approach. CE includes resource efficiency, waste and emission reduction, as well as extending product use, recycling, and repurposing materials. This chapter examines how circular economy can help manufacturing through eco-design, remanufacturing, recycling, and closed-loop production systems. It shows how business can benefit from circular economy (CE) principle and incur cost savings and regulatory compliance. Also, the chapter discusses real-world case studies where manufacturers have adopted circular practices successfully showcasing both opportunities and challenges in the perspective of transition. By using circular economy strategies, industries can lower their environmental footprint while enhancing resource efficiency and creating a more future-ready manufacturing ecosystem.

Chapter 2 highlights the revolution of Industry 5.0, the synthesis of human–machine collaboration, and prediction analysis. Conventional recycling systems are commonly plagued by manual sorting inefficiencies, nonexistent real-time monitoring, and insensitivity to change in waste composition. Industry 5.0 is centered on human-centric automation, which enables smart and efficient waste management. This chapter considers current research work on Industry 5.0 applications in recycling, human–machine interaction, and predictive analysis. Future robotics is discussed in this chapter.

AI-based waste sorting and predictive maintenance in recycling centers identify how this chapter helps in addressing sustainability. Human–machine interaction is shown to increase the accuracy of waste sorting, contaminates less, and increases recycling levels. Human expertise and advanced robotics improve decision-making and responsiveness. Predictive analysis, through AI and machine learning, optimizes waste collection, forecasts the rate of material recovery, and lowers operational costs through minimizing equipment failures. Thus, the chapter concludes that Industry 5.0 facilitates a greener recycling process through the implementation of smart automation and human smartness. Combining the two maximizes efficiency, minimizes adverse effects on the environment, and maintains circular economy goals. Future research would need to focus on scaling up AI usage and developing policies for implementing Industry 5.0 technology in waste management on a large scale.

Chapter 3 highlights waste management for sustainability and a circular economy. Traditional models are not sufficient to solve the problem of waste management. In this scenario, new technology like artificial intelligence (AI) can solve this complex problem in manufacturing companies. This research will analyze the role of AI in waste management, including assembly, arrangement, reprocessing, and controlling. It demarcates the possible advantages and problems of using AI and highlights the enhanced quality of information, confidentiality, reduction of cost, and moral deliberations. AI holds a substantial role in effective waste management by solving issues like confidentiality concerns, data quality, and reduction of cost. Optimum utilization of resources and sustainability can be achieved through AI-driven waste management systems. This chapter connects the information gap and simplifies the understanding of the use of artificial intelligence for waste management. It will support budding research, policymakers, governments, municipalities, and also organizations dealing with waste management.

Chapter 4 talks about AI-driven waste minimization model in manufacturing. The study investigates the potential use of AI-based waste minimization models in manufacturing by reviewing literature on how they have been used and identified gaps to the research on this topic. The goal of the presentation is to synthesize the studies for knowledge synthesis on applications of AI in waste reduction and formulate areas of relevant future work. The methodology of a systematic literature review was used to analyze the peer-reviewed studies retrieved from the reputable databases. The focus of the data analysis was on key AI models including machine learning (ML), deep learning (DL), and predictive analytics, and the extent to which they have contained to save on waste in different manufacturing sectors. As a result, the findings show that in process optimization, defect detection, and predictive maintenance, AI models contribute significantly to reducing material waste, energy consumption, and overproduction.

Chapter 5 examines the implementation of circular economy principles in the construction industry. As one of the world's biggest resource users, the construction industry remains among the leading sectors generating waste emissions at a global scale. The circular economy (CE) converts into an alternative for sustainable development through measures which optimize resource use alongside minimizing waste and promoting material recycling. This chapter investigates the successful implementation of circular economy principles by Skanska which operates as an international construction industry leader. The analysis contains an evaluation of Skanska's strategic approaches together with performance challenges alongside quantitative assessments that represent the company's management of waste generation while reducing carbon emissions and material waste. The main research results confirm that innovative approaches together with stakeholder participation and supportive governmental policies lead to successful circular management in construction activities.

Chapter 6 discusses about an enhanced imbalanced garbage classification model using attention-based ResNet50 with generative adversarial networks (GANs). The processing of waste faces significant difficulties from class imbalance because training datasets contain insufficient examples from specific classification categories. A proposed enhancement of ResNet50 includes deep attention mechanisms as well as GAN-based synthetic data augmentation. ResNet50 combined with attention modules integrated to a GAN model which produces synthetic minority class samples generates superior results in garbage classification systems. This model produces a better accuracy of 92.1% along with enhanced F1-score against standard ResNet50 when processing imbalanced datasets according to experimental outcomes.

Chapter 7 explores AI-powered waste management. Waste management has become a worldwide problem that requires creative ideas to maintain sustainability and create a circular economy (All noman, 2022). Conventional models are regularly failing to address the growing waste levels and their effects on ecology. In this era, traditional models are to be converted into modern technologies for effectively tackling waste. Here, the focus will be on automated sorting, recycling enhancement, predictive analytics, and ethical issues. The use of artificial intelligence (AI) in waste management is also highlighted. It investigates practical applications, including robotics combined with machine learning (ML) and AI-powered optical sorting devices for effective e-waste management. This also focuses on case studies based on the Internet of Things (IoT) and machine technologies in optimizing operations, lowering costs, and minimizing environmental risk. The chapter also closely reviews ethical issues and policy frameworks, influencing waste management AI acceptance. This study offers researchers, policymakers, and companies to work together to bridge the knowledge gap so they may cooperatively improve sustainable waste management techniques. These developments offer a radical change toward a future based on ecologically robust resources.

Chapter 8 titled "Intelligent Systems in Waste Management: A Machine Learning Approach" examine the revolutionary ability of machine learning (ML) to revolutionize waste management systems. Classical waste management methods, which are often characterized by inefficiencies, labor, and high operational expenditure, are becoming more and more supplemented or even replaced by data-driven, smart solutions. The chapter discusses how various machine learning methods, including supervised learning, unsupervised learning, and reinforcement learning, are applied to automate garbage sorting, forecast waste generation, optimize collection routes, and enhance recycling efficiency. Real-world case studies illustrate the application of ML in enhancing operating efficiency, reducing costs, and promoting sustainability in urban garbage management systems. However, evident the advantages may be, drawbacks such as quality of data, high initial outlay, and change resistance are debated. To conclude, the chapter peeks into the future, where smart cities and circular economy models interact with machine learning, and even the ethical concern of applying the intelligent systems comes into focus. This chapter discusses a comprehensive overview of the current state and future potential of machine learning applied to waste management, highlighting the promise of improving both sustainability and efficiency in the waste management sector.

Chapter 9 explores bridging corporate social responsibility and circular economy in manufacturing in light of economic, technological, and policy barriers. The corporate social responsibility (CSR) and circular economy (CE) practices are prerequisites for the promotion of sustainability in the manufacturing sector. While CSR ensures ethical and responsible business, CE emphasizes efficiency in the use of resources, minimization of waste, and closed-loop production. Despite their common goals of sustainability, the integrated role of CSR and CE in manufacturing needs to be explored to uncover their potential in larger phenomena. This chapter discusses the current literature on CSR and CE, highlights key research gaps, and presents directions for future research. Some of the key challenges are poor stakeholder engagement, economic and technological limitations, regulatory and policy hurdles, insufficient integration between CSR and CE strategies, and the lack of a standard assessment tool. Closing these gaps calls for interdisciplinary research, policy measures, and strategic frameworks that facilitate CSR-led CE adoption.

Chapter 10 examine sustainable urban air mobility and waste managemnt. Urban air mobility (UAM) is rapidly emerging as a promising solution to alleviate urban congestion, reduce transportation time, and enhance the sustainability of metropolitan environments and ensuring proper waste management practices for a healthy living environment. However, integrating UAM into existing urban infrastructures presents complex challenges across technological, regulatory, safety, environmental, societal dimensions, and waste management. This research aims to provide a comprehensive analysis of the key obstacles hindering the large-scale deployment of UAM, including technological limitations of electric vertical take-off and landing (eVTOL) aircraft, airspace congestion, regulatory uncertainties, and public acceptance. The chapter delves into the current state of eVTOL development, examining issues related to infrastructure, regulatory framework, safety standards, and system reliability. UAM involves aerial transportation systems such as drones and eVTOL aircrafts which can also in turn help cities in waste management in urban as well as rural areas. UAM is not just for passenger transportation with ease but also has its contribution toward public service operations such as waste collection and disaster response. But inculcating UAM for the use of waste management practices has its own challenges which India needs to overcome. Through case studies and various references, the chapter identifies the potential challenge initiatives aimed at addressing these barriers, such as simulation-based air traffic management systems and impact of UAM in waste disposal methods. The research concludes by offering recommendations for overcoming these challenges, emphasizing the need of public-private partnerships, and long-term sustainability in UAM

Chapter 11 talks about sustainable supply chain management (SSCM) for businesses that are trying to combine economic, environmental, and social performances. The sustainability and efficiency of supply chain are significantly enhanced by inclusion of smart technologies like the artificial intelligence (AI), the Internet of Things (IoT), blockchain, and big data analytics. Through the automation of decision-making, real-time tracking and updates, coupled with predictive analytics, smart technology enables improved productivity of resources, reduction in waste, and boost in transparency. AI-powered predictive models support reduced inventory costs, avert overproduction, and estimate demand fluctuations accurately resulting in a positive environmental impact. Through traceability and end-to-end visibility enabled by the utilization of IoT devices, real-time tracking of energy usage during storage and transportation becomes a reality. Blockchain technology contributes in ensuring data integrity and traceability, thereby increasing stakeholder accountability at one end and confidence at the other.

Chapter 12 investigates the role of technology in sustainable logistics and inventory management with reference to Food Corporation of India. Agriculture accounts for 18.2% of India's GDP and more than 450 million people are employed in the sector. Despite obstacles, supply chain integration (SCI) and developments in the agrifood industries are changing. SCI improves performance and guarantees food quality in the intricate global supply chain. SCI is converting food supply chains into incredibly effective digital systems by combining them with technological innovations like blockchain, artificial intelligence (AI), and the Internet of Things (IoT). AI is being integrated by the Food Corporation of India (FCI) to improve distribution, storage, procurement, and logistics. It helps increase control, coherence, and operational efficiency while tackling waste, delays, and inefficiencies. Furthermore, AI in agriculture is anticipated to transform food security and service delivery for stakeholders. This chapter explores different technological aspects that FCI uses and gives policy suggestions to improve the various processes.

Chapter 13 explores selling generic food items versus organic food items for health benefits. It is like learning science; the more you know, the better you will be with the application. Similarly, understanding what a consumer is looking for in sustainable food and knowing how to profile the product for marketers is crucial for product success in developing economies. This assumption served as the foundation for the current study's goal of addressing the gap left by the previous research. Among these are the food values that consumers place the highest priority on when making purchases of sustainable food items. To find a variety of food values in the pursuit of sustainable food orientation, this study divided the consumers into three customer profiles. The Lusk & Briggeman food dimensions were used in this study to gauge consumers' intentions to buy sustainable foods in underdeveloped nations.

Chapter 14 explores about migrant entrepreneurs who have overcome all the difficulties and made their venture a big success. This chapter explores the understudied topic of immigrant entrepreneurs and how these immigrants have created a viable business venture employing firm capabilities and intellectual capital that resulted in the venture enterprises' successful expansion over a period of six decades. The chapter uses an original interpretative case-based technique to convey a thorough understanding of the many difficulties experienced by migrant entrepreneurs and how they overcame them. The chapter also highlights many transitions and changes that occurred in the six decades following India's partition by focusing on one of the rising industries in India, the watch business. The chapter fills a gap in the literature about sustainable business ventures started by immigrant entrepreneurs in a developing nations and how the entrepreneur overcome different challenges and brought sustainability to the business and the ecosystem by not just efficiently using the resources but also reducing the waste that is generated through scrappage.

Waste management is an important global challenge, where increasing urban ization and industrialization lead to high waste production. Traditional waste landfill methods, including manual sorting and fixed collection plans often result in disabilities, increased costs, and environmental decline. Integration of machine learning (ML) and artificial intelligence (AI) in waste manage ment systems provides a transformative approach to solve these problems. This chapter examines the role of ML in automatic waste sorting, future indication analysis, collection passport optimization, and converting waste to-energy. Case studies from Singapore, San Francisco, and large waste Chapter 15 explores role of machine learning in waste management. management companies highlight ML's real-world applications to create smart and more sustainable waste management systems. Singapore's AI operated system and San Francisco's policy-based waste management models show ML's ability to adapt to comparative analysis resources and reduce landfill addiction. The findings show that AI-controlled waste management leads to high recycling speeds, reduces greenhouse gas emissions, and more efficient resource allocation.

List of Figures

List of Tables

List of Contributors

Agrawal, Priyanka, *Graphic Era deemed To be University, India*

Ahuja, Mamta, *Department of Commerce, Graphic Era Deemed To be University, India*

Arora, Madhu, *Symbiosis Institute of Computer Studies and Research, Symbiosis International (Deemed University), India*

Arora, Yugank, *Blue Heavens, Australia*

Arul, N., *Department of CSE (AIML), St. Peter's Engineering College, India*

Bhandari, Kalpana, *H.N.B. Garhwal University, India*

Bhatt, Avani, *Department of Commerce, Graphic Era Deemed To be University, India*

Choudhury, Madhusmita, *Assistant Professor, Department of Marketing, GITAM deemed To be university, India*

Damodharan, V. S., *Department Chair Business, Abu Dhabi Vocational Education & Training Institute, United Arab Emirates*

Ghai, Sachin, *Department of Management Studies, Graphic Era Deemed To be University, India*

Goel, Richa, *Associate Professor, Graphic Era University, India; Symbiosis Centre for Management Studies, Symbiosis International Deemed University, India*

Govindapuram, Suresh, *IFMR GSB, Krea University, India*

Jena, Swarna Prabha, *Department of Electronics and Communication Engineering, Centurion University of Technology and Management, India*

Kathuria, Samta, *Parul Institute of Law, Parul University, India*

Khan, Shafqat, *Parul Institute of Law, Parul University, India*

Khanna, Rupa, *Professor, Graphic Era University, India*

Kishan M., Kaneriya, *Assistant Professor, Vanbandhu Polytechnic in Agriculture, S.D.A.U., India*

Krishna E S, Anuram, *Department of Global Financé Analytics, Christ University, India*

Kumar, Rakesh, *Department of Management Studies, Graphic Era Deemed To be University, India*

Kumari, Arpana, *Assistant Professor, Symbiosis Centre for Management Studies, Symbiosis International Deemed University, India*

Kumari, Prity, *Assistant Professor, College of Horticulture, A.A.U., India*

Latiyan, Anshu, *Department of Management Studies, Graphic Era Deemed To be University, India*

M, Ramasubramaniam, *Area Chair – Business Analytics Professor of Operations & Analytics, Loyola Institute of Business Administration (LIBA), India*

Mishra, Swetalina, *School of Management, Centurion University of Technology and Management, India*

N, Chandrasekaran, *IFMR GSB, Krea University, India*

Nandini, Girija, *School of Management, Centurion University of Technology and Management, India*

Nanjundan, Preethi, *Department of Data Science, Christ University, India*

Paikaray, Bijay Kumar, *Centre for Data Science, Department of Computer Science and Engineering, Siksha 'O' Anusandhan (Deemed to be) University, India*

Raj, Shantanu, *Centre of AgriManagement, Utkal University, India*

Rifai, Firas, *Al-Zaytoonah University of Jordan, Jordan*

S. J., Devra, *Ph.D. Scholar, Department of Agricultural Statistics, College of Agriculture, J.A.U., India*

S. R., Raj, *Senior Research Fellow, Department of Agricultural Statistics, College of Agriculture, A.A.U., India*

Samal, AdityaRanjan, *Centre of AgriManagement, Utkal University, India*

Sharma, Manu, *Department of Managememt Studies, Graphic Era Deemed To be University, India*

Sharma, Neeraj, *Graphic Era Deemed To be University, India*

Sharma, Priti, *Department of Commerce, Graphic Era Deemed To be University, India*

Sharma, Rajat, *Entrepreneurship Development Institute of India-Ahmedabad*

Shaw, Priyal, *Parul Institute of Law, Parul University, India*

Singh, Muskan, *Graphic Era Deemed To be University, India*

Singh, Shruti, *Parul Institute of Law, Parul University, India*

Thomas, Lijo, *Department of Psychology, Christ University, India*

Vidya, R., *Department of CSE (AIML), St. Peter's Engineering College, India*

List of Abbreviations

AAM	Advanced air mobility
AC-GAN	Auxiliary classifier GAN
AGA	Automatic grain analyzer
AGV	Automated guided vehicles
AI	Artificial intelligence
ANN	Artificial neural network
ANOVA	Analysis of variance
AR	Augmented reality
ARIMA	Autoregressive integrated moving average
ATC	Air traffic control
BAIL	Bengaluru International Airport
BIM	Building Information Modeling
CAP	Cover and plinth
CAGR	Compound annual growth rate
CBAM	Convolutional block attention modules
CCEA	Cabinet Committee on Economic Affairs
CE	Circular economy
CFPP	Centralized Food Grain Procurement Portal
CII	Confederation of Indian Industry
CLS	Closed-loop system
CM	Circular manufacturing
CNN	Convolutional neural network
CORUS-XUAM	Concept of operations for European
UTM systems	Extension for urban air mobility
CPS	Cyber-physical systems
CSR	Corporate social responsibility
cGAN	Conditional GAN
DFPD	Department of Food and Public Distribution
DGCA	Directorate General of Civil Aviation
DGGAN	Deep convolutional GANs
DL	Deep learning

DLW	District-level warehouse
DOS	Depot online system
DTC	Direct to consumer
DSNY	New York City Department of Sanitation
EASA	The European Union Aviation Safety Agency
EHS	Environment health and safety
ERP	Enterprise resource planning
EO	Entrepreneurship orientation
EPR	Extended producer responsibility
FAA	Federal Aviation Administration
FCI	Food Corporation of India
FPS	Fair price shop
FSC	Food supply chain
GA	Genetic algorithm
GAN	Generative adversarial network
GDP	Gross domestic product
GDPR	General Data Protection Regulation
GHG	Global greenhouse gases
GPP	Green public procurement
HC	Human capital
HMT	Hindustan Machine Tools
HVAC	Heating, ventilation, and air conditioning
IC	Intellectual capital
I4.0	Industry 4.0
I5.0	Industry 5.0
IESA	India Energy Storage Alliance
IMC	Integrated marketing communication
IoT	Internet of Things
ISO	International Standards Organization
IT	Information technology
K-UAM	Korean Urban Air Mobility
LAA	Los Angeles Airways
LCA	Life cycle assessment
LED	Light emitting diode
LR	Linear regression
LPG	Liberalization, privatization, and globalization
LSTM	Long short-term memory
ML	Machine learning
MoU	Memorandum of understanding

MRF	Material recycling function
MSP	Minimum support price
MSW	Municipal stable waste
NASA	National Aeronautics and Space Administration
NDARC	NASA Design and Analysis of Rotorcraft
NEA	National Environment Agency
NFSA	National Food Security Act
NGO	Nongovernmental organization
PaaS	Product-as-a-service
PDS	Public distribution system
PIMS	Procurement information management system
PPP	Public-private partnership
RBV	Resource-based view
RC	Relational capital
RNN	Recurrent nerve network
ROI	Return on investment
SC	Supply chain
SCI	Supply chain integration
SCM	Supply chain management
SDG	Sustainable development goals
SDG	Sustainable Development Goal
SESAR	Single European Sky ATM Research
SLR	Systematic literature review
SME	Small and medium enterprise
SSC	Sustainable supply chain
SSCM	Sustainable supply chain management
SUAM	Sustainable urban air mobility
SWM	Solid waste management
SVR	Support vector region
SVM	Support vector machine
TIA	Type inspection authorization
ToU	Time-of-use
TOI	Times of India
TPB	Theory of planned behavior
UAM	Urban air mobility
UAV	Unmanned aerial vehicle
UN	United Nations
UTM	Unmanned traffic management
VBN	Value-belief norm

VLTS	Vehicle location tracking system
VR	Virtual reality
WINGS	Warehouse inventory network and governing system
WTE	Waste-to-energy
XAI	Explainable AI
Yolo	You only see once
ZWIA	Zero Waste International Alliance

1

Introduction to a Circular Economy and its Applications in the Manufacturing Process

Anshu Latiyan and Sachin Ghai

Department of Management Studies, Graphic Era Deemed to be University, India
E-mail: anshulatiyan.mgt@geu.ac.in; professorsachinghai@gmail.com

Abstract

At present, due to fast-changing industrial scenario, companies are in search of economic development and environmental balance. The circular economy (CE) represents an alternative to the traditional take-make-dispose approach. CE includes resource efficiency, waste and emission reduction as well as extending product use and recycling and repurposing materials. This chapter examines how a circular economy can help manufacturing through eco-design, remanufacturing, recycling, and closed-loop production systems. It shows how businesses can benefit from the circular economy (CE) principles and incur cost savings and regulatory compliance. Also, the chapter discusses real-world case studies where manufacturers have adopted circular practices successfully, showcasing both opportunities and challenges in the perspective of transition. By using a circular economy strategies, industries can lower their environmental footprint while enhancing resource efficiency and creating a more future-ready manufacturing ecosystem.

Keywords: A circular economy, sustainable manufacturing, resource efficiency, waste reduction, green manufacturing, circular supply chains.

1.1 Introduction

It is crucial to take into account the different challenges that rivals must face in order to compete in the cutthroat, technologically sophisticated, and

DOI: 10.1201/9788743806905-1

ever-changing business environment of today. The circular economy concept has gained international attention due to problems with waste pollution and resource depletion brought on by the global reach of corporate operations (Uwuigbe et al., 2025). Commodity producers are now facing significant constraints owing to the lack of natural resources and unequalled ecological degradation therefore, the ground realities of economics call for a different approach to the implementation of economic activity. A new paradigm of development is needed today for sustainable growth that is based on economic growth, protection of the environment, and well-being of mankind. The linear economic model of many nations is predicated on the "take, make, dispose" premise. Because of this, the conventional economic model is being rethought, and the idea of a circular economy – which is founded on the "take, make, reuse" principle – is receiving increasing attention (Tambovceva and Titko, 2020). The idea of the circular economy (CE) is becoming more popular among academics, business leaders, legislators, and practitioners who have characterized it as an alternate paradigm to reduce emissions, waste, and resource depletion (Su et al., 2013). A Circular economy is characterized as a sustainable development strategy focused on resource optimization and waste reduction (Haupt and Zschokke, 2017). The approach examines product design and encourages reuse, repair, remanufacturing, and recycling to create a closed-loop system that minimizes resource input and waste creation, in contrast to the traditional linear economy model. Recently, businesses and decision-makers are beginning to realize the importance of discarding the conventional linear economic model of "take, make, dispose." Hence, a circular economy (CE) is a practical alternative that focuses on waste reduction by extending the life of resources and products while reducing activities that produce waste.

Transitioning to a circular economy offers numerous long-term benefits; however, it does have its downsides. (1) The difficulty of implementing system changes is tough. (2) Financial problems arise when a circular economy cannot be profitable in the short term. (3) Insufficient markets may occur due to lack of competition, knowledge, incentives, physical infrastructure, or essential products in an area. (4) Imperfect control (rules and/or execution are insufficient). (5) Problems in society like an absence of knowledge related to a circular economy, and skills. (6) Problem to get enough money. (7) Lack of uniform processes in different areas. Basically, the CE is an economic concept aimed at establishing a closed-loop production cycle in which materials are continuously recycled or repurposed, waste is reduced, and resources are employed effectively. To guarantee that materials stay in circulation for as long as possible, CE promotes reuse, recycling, remanufacturing, and responsible consumption in contrast to the traditional approach, which discards things after use. Several research papers offer valuable insights into developing effective a circular economy framework.

1.1.1 Key principles of the circular economy

There are certain concepts on which the circular economy is built that guide startups and companies toward sustainability.

- **Prevention of pollution and waste**
  With keeping sustainability in mind, products and processes should be designed to ensure that there is less waste creation and at every level of production the pollution should be reduced. This includes employing eco-technologies, optimization of resource efficiency, and using biodegradable materials.
- **Maintaining the use of materials and products**
  Rather than disposing products after only one use, a circular economy promotes restore, refurbishment, and rebuilding so that the lifespan of products can be extended.
- **Natural system regeneration**
  The main goal of a circular economy is not only to reduce environmental damage but also to emphasize the restoration and improvement of ecological systems. It incorporates waste management, organic farming, and sustainable energy implementation. By combining these assumptions, industries can transition to a more strong and sustainable financial system, leading to long-term environmental and economic benefits.

1.2 Shift from Linear to Circular Frameworks

The linear economy, also known as the take-make-waste economy, is an approach in which resources are collected to create products that are eliminated as garbage. Conventional economic models and structures that utilize an approach to both the consumption and manufacturing are known as linear economy (Baranski, 2021).

1.2.1 Linear economy limitation and criticisms:

- **Feedback absence:**
  As the name itself suggests, linear economy differs from a circular economy because it does not have a feedback system. Firms from a

range of sectors and businesses contribute to the economy by providing the services and the goods with an aim of earning profit by selling and distributing them to the potential consumers.

- **Negative impact on the environment:**
  The linear economy model has a great impact on the environment due to its heavy reliance on resource extraction and mass production. Firms extract natural resources to meet the growing demand for consumer goods even at an unsupportable rate. For example, the production of consumer electronics depends on the mining of rare earth elements, which are essential for components like screens, circuit boards, and batteries.

1.2.2 The circular economy approach

Economic activity creates and restores the general health of the system in a circular economy. The idea acknowledges how crucial it is for the economy to function well at all levels, for both large and small enterprises, organizations, and individuals, and on both local and global scales (Macarthur and Heading, 2019). Not only it reduces waste, it also aims to restore natural ecosystem by reducing pollution and encouraging sustainable resource management.

1.2.2.1 It is based on three principles:

- Reduce waste and prevent pollution.
- Maximize the lifespan of products and materials.
- Restore and enhance natural system

1.3 Literature Review

Numerous concepts and theoretical frameworks serve as the foundation for the circular economy (CE) concept. This idea has evolved over the last few decades by combining many theories and methods to enhance resource efficiency and environmental sustainability (Yuan and Moriguichi, 2006), (Chauhan et al., 2022).

The number of publications on the circular economy has increased significantly in recent years, reflecting growing academic and industry interest. Sustainability, waste management, and creative business models are among the major themes that have been found, underscoring initiatives to overcome implementation obstacles (Uwuigbe et al., 2025).

A growing number of people view the circular economy (CE) concept as a way to address a number of issues, including waste production, resource scarcity, and maintaining financial gains. However, the notion of circularity is not particularly new. In the past, certain situations and incentives have sparked concepts related to circularity through recycling, remanufacturing, and reuse (Sondhi et al., 2024).

Our economy has been linear since the Industrial Revolution. The world has become a "take, make, dispose" environment due to our consumer and "single use" lives. This alludes to a unidirectional production paradigm in which our industrial inputs are derived from natural resources and utilized to manufacture mass-produced commodities that are bought and, usually, thrown away after only one usage. The physical boundaries of the world are being tested by this linear economy model of mass production and consumption. As a result, it cannot continue, and a shift to a circular economy is becoming increasingly unavoidable (Esposito et al., 2018).

1.4 Application of a Circular Economy in Manufacturing

- **Design for longevity**
  Increasingly, manufacturers are concentrating on creating long-lasting, easily maintained, and repairable products. Businesses can ensure that their products last longer and need fewer replacements by utilizing premium materials and modular designs. For instance, to prolong the useful life of their products and reduce the need for frequent disposal, appliance manufacturers now provide maintenance instructions and spare parts.
- **Resource efficiency**
  Costs can be decreased, and the environmental impact can be minimized by making the most efficient use of energy and materials throughout the production process. To maximize resource use, numerous industries are implementing lean production methodologies, sophisticated automation, and energy-efficient technologies. For example, to reduce energy usage during production, some automakers increasingly use lightweight, recyclable materials.
- **Minimization of waste**
  The main principle of CE is to reduce wastage. Several companies and organizations are shifting to zero-waste production models. For example, to ensure minimum wastage, multiple textile firms are turning fabric scraps into insulating materials.

- **Product life extension**
  Remanufacturing, repairing, and repurposing things to extend their lifespan contribute to keeping materials in circulation. To reduce electronic waste, electronics businesses, for instance, are offering trade-in programs where outdated gadgets are repaired and resold. Similarly, in an effort to reduce resource depletion, the furniture business is pushing reconditioned or repurposed goods.
- **Closed-loop system**
  The motive of a CLS is to continuously cycle waste back into production rather than into landfills by emphasizing recycling and material reuse. By developing goods with recyclability in mind, several sectors are ensuring that materials may be recovered and used in new production cycles at the end of their lifecycle. The substitution of recycled plastics for virgin plastics in packaging is one example of this.

1.5 Benefits of a Circular Economy into Manufacturing

Although the potential of the circular economy (CE) in manufacturing processes has been recognized, the concept of circular manufacturing (CM) and its implementation are not clearly stated and can sometimes be confused with the more general idea of sustainability (Acerbi and Taisch, 2020). Adopting the circular economy in manufacturing provides various benefits, such as reducing costs through reuse, increasing innovation, and enhancing environmental sustainability. It also creates employment opportunities and further boosts brand recognition.

- **Environmental benefits:** There is a decrease in waste and resource extraction, which reduces the impact on the environment. Undoubtedly, incorporating a circular economy (CE) offers various benefits for the implementer's environmental approach (Quarterly, 2017). The approach should begin at the micro level with enterprises to effectively implement the circular economy. By supporting waste reduction and reducing the mining of raw materials, the circular economy significantly decreases the burden on natural resources. Enterprises can minimize pollution and carbon footprints by focusing more on remanufacturing, recycling, and environmentally friendly resource management. When recycled metals are used in production instead of virgin ores, energy consumption is reduced, which reduces environmental degradation.

- **Economic benefits:** A circular economy in manufacturing helps cut costs by reusing materials, improving resource efficiency, and providing new employment opportunities. Industries can repurpose and reclaim materials instead of discarding them, which will definitely reduce production costs. In addition, circular business models provide competitive advantages and new sources of income in sectors such as product-as-a-service, repair, and refurbishment. Companies that sell remanufactured devices, for instance, can satisfy budget-conscious customers while still generating a profit.
- **Social benefits:** Making the switch to a circular economy helps in creating job opportunities, from recycling and remanufacturing to sustainable product design. These and other such related sectors are seeing an increased demand for skilled workers in waste management, materials recovery, and repair services. These changes provide further resilience to the economy of the region, where sustainable industries can guarantee stable employment opportunities and foster local economic development.

1.6 A Circular Economy Opportunities

Various social and political options are available in a circular economy. In simple terms, it enhances the relationship between industry and society. Closing the loop requires long-term cooperation from all supply chain performers, including the general public and organizations (Geng et al., 2012). Since the value chain no longer ends with consumers, a product should be recovered at the end of its life. Businesses and customers are better aligned as a result of this positioning, because of which the organizations can understand the expectations and demands of consumers in a more suitable way and supply them goods as per their expectations or requirement. This will result in satisfying the customers and will attract many more customers. People are now focusing more and being mindful when choosing safe and environmentally friendly products. Additionally, the circular economy provides rental models across all businesses (MacAthur, 2013) that help in assisting businesses in collecting customer insight data and offering more individualized and customized products at minimal costs in accordance with customers' needs and desires. As a result, life quality improves and social values are promoted. A circular economy majorly focuses businesses to follow the rules and regulations.

CE increases businesses' environmental awareness and establishes the validity of the organization. For instance, they can lessen societal pressure and comply with legal obligations. CE also helps businesses increase their profitability and save money. It makes it possible to cut expenses by implementing end-of-life and sustainable supply chain management, lowering input costs, and minimizing waste production and environmental penalties (Park et al., 2010) (Geng et al., 2012). Businesses can generate more revenue by selling their waste rather than throwing it away, thanks to the closed-loop supply chain model. Additionally, CE creates new markets for remanufacturing and recycling (Kumar et al., 2019).

The new market and source of income offer established businesses a competitive edge over their competitors and maximize their revenue. Innovative concepts that aim to safeguard the environment have developed as a result of advancements in environmental sciences and technologies, including eco-design, eco-label, cleaner production, and life cycle assessment (Geng and Doberstein, 2008). Therefore, the rise in green products and environmentally friendly goods has led to their wide usage and acceptance, as they have a minimal impact on the environment.

1.7 Challenges to a Circular Economy

- **Economic:** One of the biggest hurdles in adopting a circular economy is the high initial investment required for new technologies and infrastructure. There are many businesses that hesitate to shift from traditional linear models because the financial returns are not always immediate. In addition, consumers often rely on references and reviews when deciding to adopt new products.
- **Technological:** For a circular economy to work effectively, industries need advanced recycling and remanufacturing technologies. Although it is difficult to reuse and recover many materials that are used in manufacturing, products with complex designs, such as automobiles and electronics, need innovative solutions to separate and process materials without compromising quality. To develop such technologies is time-consuming and needs resources that could slow down the process.
- **Regulatory:** The lack of industry standards and clear policies makes it difficult for businesses to embrace a circular economy practices. In many areas, there are fewer incentives or legal requirements that push the companies to reduce waste and focus more on the reuse of materials.

There are various businesses that are opposed to making the shift without tax benefits, strong government support, or subsidies. Regulatory and policy-related hurdles can provide significant challenges. To overcome these challenges and make sustainability an achievable and profitable choice for businesses, organizations, customers, and lawmakers must work collaboratively.

1.8 Barriers to Implementing a Circular Economy in Manufacturing

Although the circular economy (CE) provides various opportunities, the degree of public awareness regarding CE is very impecunious, as there are many people who are not aware of the CE concept. This is one of the hurdles in implementing the circular economy into manufacturing. Despite the reality that governments and businesses globally have been engaging in CE initiatives in previous years, the idea and principles of CE are still not widely known (Benton et al., 2017). There are various economic barriers to the a circular economy in the manufacturing sector, as a circular economy is very expensive and it requires a significantly higher amount of investment (Liu and Bai, 2014). According to the research papers, prior to making a purchase, the majority of people are primarily concerned with product design. People select the product that appears nicer rather than the one that is made from waste and often overlook sustainability or the consequences for the environment (Pomponi and Moncaster, 2017), (Naustdalslid, 2014).

A circular economy also faces environmental hurdles, which include a lack of sustainability development programs and initiatives at educational institutions and government organizations, as well as the inefficiencies that presently exist. The incentives currently offered to encourage environmentally friendly practices and resource conservation are not enough to drive beneficial change. Although initiatives to conserve water, materials, and energy do exist, they usually fall short of what is needed to have a meaningful effect. To fully implement sustainable practices, companies and sectors need stronger monetary support and technological development. The move towards a greener environment is still challenging and delayed due to the absence of adequate incentives (Geng et al., 2009).

- **High initial investment cost** Shifting to a circular model often requires significant financial expenditures in new technologies, infrastructure, and employee training. Various manufacturing firms, majorly small and medium-sized enterprises (SMEs), are unwilling to implement sustainable management practices in the absence of financial benefits.

- **Limited technological advancements**
  Some materials and goods are still challenging to recover and reuse effectively, even with advancements in recycling and remanufacturing technologies. Many items were not designed with easy disassembly or recycling in mind, which results in higher costs and inefficient material recovery. To overcome this obstacle, innovation in production methods and material science is crucial.
- **Regulatory and policy gaps**
  The adoption of circular practices is frequently slowed down by inconsistent policies and government incentives. There are few legal or financial incentives for companies to invest in sustainable alternatives, and policies in many nations continue to support linear production methods. Industries are not particularly motivated to change their operations in the absence of clear legislation that supports circular practices.
- **Supply chain complexity**
  Coordination between different stakeholders, such as manufacturers, recycling companies, and suppliers, is necessary to implement a circular economy. The "take-make-dispose" architecture of many current supply chains makes it difficult to incorporate closed-loop operations like resource recovery, remanufacturing, and product refurbishing.
- **Market and consumer behavior**
  The success of CE in manufacturing is largely dependent on consumer preferences and market demand. Many consumers continue to favor new products because they believe they are more dependable and of higher quality than reconditioned or recycled ones. For broader acceptability, it is imperative to alter consumer perception and raise awareness of the advantages of circular products.

1.9 Future Directions

The circular economy is starting to gain popularity as industries realize how important it is to manage resources sustainably. Technological developments, creative corporate strategies, and governmental regulations targeted at reducing waste and maximizing resource utilization will all influence the circular economy in manufacturing in the future.

1.9.1 Emerging trends:

1.9.1.1 Technologies and business models supporting CE

In instance, to effectively use the concepts of the circular economy, manufacturers are adopting new technologies and approaches among the major trends influencing the future. Usually, moving from linear economy to a circular economy frameworks demands major initial investments in training, infrastructure, and technology. It could be very difficult to ensure that these investments generate long-term economic benefits, specifically, for small and medium enterprises with limited capital.

- **Smart manufacturing and Industry 4.0**:
  The integration of Internet of Things (IoT), AI, and data analytics allow businesses to monitor material usage, enhance efficiency, and cut waste through automation and predictive maintenance.
- **Recycling innovations**
  The capacity to recover and reuse materials, especially from intricate goods like electronics and plastics, is being enhanced by developments in mechanical and chemical recycling.
- **3D printing and additive manufacturing**: By facilitating on-demand production, these technologies lower material waste and make it possible to use recyclable resources in the production process.
- **Product-as-a-service (PaaS) models**: Businesses are moving away from outright product sales and toward leasing or subscription-based business models, which guarantee that things are maintained, reconditioned, and reused rather than thrown away.
- **Eco-friendly and biodegradable materials**: The environmental impact is being lessened by the adoption of plant-based substitutes, recycled metals, and biodegradable plastics as a result of research into sustainable materials.

1.9.2 Policy directions: Government initiatives promoting a circular economy

To support the circular economy through effective laws and regulations, governments are promoting the need for CE.

Such initiatives includes:

- **Extended producer responsibility (EPR):**
  Promoting a circular economy with the goal to ensure that their products are easily fixed, recycled, or reused, producers are being held accountable for the full lifecycle of their products.

- **Financial incentives for sustainability:**
  Circular businesses are being motivated to engage in environmentally friendly innovations and circular economic models through tax reductions, subsidies, and loans.
- **Legislation on waste reduction**
  Bans are being imposed by the government on single-use plastics. To encourage businesses to adopt more environmentally friendly procedures, waste reduction goals and stronger recycling regulations are being enforced by the government.
- **International cooperation for circular trade:**
  To initiate worldwide sustainability standards, coordinated global efforts are required to ensure consistency and effectiveness in implementation.
- **Public-private collaborations:**
  To promote a circular economy, governments are collaborating with businesses and academic institutions to support and promote advancements in waste management, recycling resources, and production techniques.

1.10 Findings

- **Increased effectiveness of resources:**
  The adoption of a circular economy (CE) practices in manufacturing leads to improved utilization of resources, reducing material waste and energy consumption. This allows industries to optimize production processes while minimizing environmental impact.
- **Reduction in waste and emissions:**
  Companies implementing circular strategies, such as recycling, remanufacturing, and closed-loop production, experience a significant reduction in industrial waste and carbon emissions, that align with the sustainability goals.
- **Economic benefits and cost savings:**
  By repurposing materials and extending product lifecycles, sustainable cost savings can be achieved by manufacturers on production expenses and raw materials, making circular economy a financially viable alternative to traditional linear production models.
- **Regulatory compliance and competitive advantages:**
  By implementing circular practices, businesses can keep ahead of environmental rules, decreasing the risk of violations and improving their position in the marketplace and brand reputation.

- **Challenges and opportunities in implementation:**
  Considering the various advantages of a circular economy (CE), industries often face difficulties like supply chain realignment, high initial investments, and upgrading technology. Successful case studies still indicate that overcoming these obstacles can result in sustainability and long-term profitability.
- **Innovation and eco-design:**
  The idea of the circular economy inspires industries to concentrate on eco-design, resulting in innovative product creation that considers sustainability from the beginning and produces a manufacturing ecosystem that is more environmentally friendly as well as prepared for the future.
- **Positive environmental and social impact:**
  Implementing a circular economy strategies not only creates employment opportunities in sustainable production sectors, recycling, and remanufacturing but also minimizes the environmental impact, helping to build a more flexible and accountable business landscape.

1.11 Conclusion

- The shift to a circular economy in manufacturing has become strategically necessary for long-term sustainability in industry, as well as an environmental requirement.
- This chapter explored how the concepts of the circular economy (CE) – such as environmental design, recycling, remanufacturing, and closed-loop production – offer substantial benefits to businesses from financial savings to adherence to regulations.
- Beyond these widely recognized benefits, though, an important fact becomes clear: the circular economy is an innovative approach that reinterprets the basic foundation of economic growth and competitiveness, not just a framework for maximizing resource utilization.
- By effectively implementing circular techniques, industries are not only eliminating waste but also transforming business models, encouraging industrial harmony, and developing new value chains that place durability over disposability.
- By prompting businesses to reconsider their dependence on virgin resources and investigate other alternatives through industrial partnerships, the shift from a linear to a circular approach encourages cross-sector interactions.

- This shift lowers dependence on uncertain international markets while also improving the resilience of supply chains and creating opportunities for regional growth in the economy.
- In addition, aligning policies and technological advancements is key to the future of circular production. The shift will be driven by emerging technologies in materials made from biomass, the use of blockchain technology for transparent substance monitoring, and artificial intelligence (AI)-driven waste management.
- Through favorable tax laws, expanded producer responsibility initiatives, and investments in environmentally conscious infrastructure, governments and policymakers can considerably support circular behaviors.
- The real achievement of a circular economy implementation will be evaluated not only by the efficacy of resources reused but also by the way industries adapt to redefine developments.
- The manufacturing industry is at a turning point in its history, and implementing circularity will not only reduce harm but also create an innovative, flexible, and adaptable economic ecosystem that utilizes sustainability as a competitive advantage.
- In the end, the circular economy intends to create new opportunities for economic innovation in addition to minimizing waste cycles.
- By implementing a circular economy principles, industries not only reduce their environmental impact but also establish themselves as innovators in environmentally friendly manufacturing by adopting the concepts of the circular economy (CE).

Acknowledgment

We are truly grateful to the faculty of the Department of Management Studies at Graphic era deemed to be University for their unwavering support and encouragement. Their guidance and the opportunities they provide have been invaluable in helping us compose this chapter.

References

Uwuigbe, U., Issah, O., Ranti, U. O., Zubeiru, M., Anaba, S., & Seidu, A. A. J. (2025). A circular Economy: A Bibliometric Review of Research in Emerging Economies (2010-2024). *International Journal of Energy Economics and Policy*, *15*(1), 77-89.

Tambovceva, T., & Titko, J. (2020). Introduction to a circular economy. *Ekonomikas un kulturas augstskola.*

Su, B., Heshmati, A., Geng, Y., & Yu, X. (2013). A review of the circular economy in China: moving from rhetoric to implementation. *Journal of cleaner production*, *42*, 215-227.

Haupt, M., & Zschokke, M. (2017). How can LCA support the circular economy?—63rd discussion forum on life cycle assessment, Zurich, Switzerland, November 30, 2016. *The International Journal of Life Cycle Assessment*, *22*, 832-837.

Baranski, A. (2021). Linear Economy: Characteristics and Criticism. *Profolus [WWW Document]. https://www.profolus.com/topics/linear-economy-characteristics-and-criticism.*

Macarthur, E. L. L. E. N., & Heading, H. E. A. D. I. N. G. (2019). How the circular economy tackles climate change. *Ellen MacArthur Found*, *1*, 1-71.

Yuan, Z., Bi, J., & Moriguichi, Y. (2006). The circular economy: A new development strategy in China. *Journal of industrial ecology*, *10*.

Chauhan, C., Parida, V., & Dhir, A. (2022). Linking a circular economy and digitalisation technologies: A systematic literature review of past achievements and future promises. *Technological Forecasting and Social Change*, *177*, 121508.

Uwuigbe, U., Issah, O., Ranti, U. O., Zubeiru, M., Anaba, S., & Seidu, A. A. J. (2025). A circular Economy: A Bibliometric Review of Research in Emerging Economies (2010-2024). *International Journal of Energy Economics and Policy*, *15*(1), 77-89.

Sondhi, S. S., Salwan, P., Niranjan, S., Hawkins, T., Behl, A., Pereira, V., & del Giudice, M. (2024). Exploring open innovation dynamics: service and delivery interplay in a circular economy. *IEEE Transactions on Engineering Management.*

Esposito, M., Tse, T., & Soufani, K. (2018). Introducing a circular economy: New thinking with new managerial and policy implications. *California Management Review*, *60*(3), 5-19.

Acerbi, F., & Taisch, M. (2020). A literature review on a circular economy adoption in the manufacturing sector. *Journal of Cleaner Production*, *273*, 123086.

Quarterly, M. (2017). Mapping the benefits of a circular economy.

Geng, Y., Fu, J., Sarkis, J., & Xue, B. (2012). Towards a national a circular economy indicator system in China: an evaluation and critical analysis. *Journal of cleaner production*, *23*(1), 216-224.

MacArthur, E. (2013). Towards the circular economy. *Journal of industrial ecology*, *2*(1), 23-44.

Park, J., Sarkis, J., & Wu, Z. (2010). Creating integrated business and environmental value within the context of China's a circular economy and ecological modernization. *Journal of Cleaner Production*, *18*(15), 1494-1501.

Geng, Y., Fu, J., Sarkis, J., & Xue, B. (2012). Towards a national a circular economy indicator system in China: an evaluation and critical analysis. *Journal of cleaner production*, *23*(1), 216-224.

Kumar, V., Sezersan, I., Garza-Reyes, J. A., Gonzalez, E. D., & Al-Shboul, M. D. A. (2019). A circular economy in the manufacturing sector: benefits, opportunities and barriers. *Management decision*, *57*(4), 1067-1086.

Geng, Y., & Doberstein, B. (2008). Developing the circular economy in China: Challenges and opportunities for achieving'leapfrog development'. *The International Journal of Sustainable Development & World Ecology*, *15*(3), 231-239.

Benton, D., Hazell, J., & Hill, J. (2017). *The guide to the circular economy: capturing value and managing material risk*. Routledge.

Liu, Y., & Bai, Y. (2014). An exploration of firms' awareness and behavior of developing a circular economy: An empirical research in China. *Resources, Conservation and Recycling*, *87*, 145-152.

Pomponi, F., & Moncaster, A. (2017). A circular economy for the built environment: A research framework. *Journal of cleaner production*, *143*, 710-718.

Naustdalslid, J. (2014). A circular economy in China–the environmental dimension of the harmonious society. *International Journal of Sustainable Development & World Ecology*, *21*(4), 303-313.

Geng, Y., Zhu, Q., Doberstein, B., & Fujita, T. (2009). Implementing China's a circular economy concept at the regional level: A review of progress in Dalian, China. *Waste Management*, *29*(2), 996-1002.

2

Tech-driven Sustainability: Smart Technologies Reshaping Waste Recycling

Priyal Shaw[1], Samta Kathuria[1], Shruti Singh[1], Shafqat Khan[1], and Kalpana Bhandari[2]

[1]Parul Institute of Law, Parul University, India
[2]H.N.B. Garhwal University, India
E-mail: priyalshaw2001@gmail.com; samtakathuria14@gmail.com; shrutisinghshru@gmail.com; shafqatk1404@gmail.com; kalpanabhandari369@gmail.com

Abstract

With the revolution of Industry 5.0, the synthesis of human–machine collaboration and prediction analysis has efficiently enhanced recycling systems, supporting sustainability. Conventional recycling systems are commonly plagued by manual sorting inefficiencies, a lack of real-time monitoring, and insensitivity to changes in waste composition. Industry 5.0 is centered on human-centric automation, enabling smart and efficient waste management. This chapter considers current research work on Industry 5.0 applications in recycling, including human–machine interaction and predictive analysis. Future robotics, AI-based waste sorting, and predictive maintenance in recycling centers are discussed to identify how this chapter helps to address sustainability. Human–machine interaction has been shown to increase the accuracy of waste sorting, reduce contamination, and improve recycling levels. Human expertise, combined with advanced robotics, improves decision-making and responsiveness. Predictive analysis, through AI and machine learning, optimizes waste collection, forecasts material recovery rate, and lowers operational costs by minimizing equipment failures. Thus, the study concludes that Industry 5.0 facilitates a greener recycling process through the implementation of smart automation and human smartness. Combining the two maximizes efficiency, minimizes adverse effects on the environment, and

DOI: 10.1201/9788743806905-2

maintains a circular economy goals. Future research would need to focus on scaling up AI usage and developing policies for implementing Industry 5.0 technologies in waste management on a large scale.

Keywords: Smart technology, Industry 5.0, artificial intelligence, sustainable manufacturing, waste management, recycling, resource optimization, circular economy.

2.1 Introduction

Industry 5.0 emphasizes the significance of human interaction in achieving environmental goals and maintaining resilience throughout the process. This is a shift toward digital and computational capabilities creating smarter and more connected workplaces (Caiazzo et al., 2022). By encouraging cloud-assisted monitoring platforms along with human centricity promotes the fundamental principle of Industry 5.0. It shifts the industry's perspective toward more sustainable and efficient recycling. Smart manufacturing systems are thereby very crucial to optimize and control the manufacturing process without really harming the environment. "Industry 5.0 will continue to grow," "trajectory would be evolutionary" (Barata and Kayser, 2023).

Integration of advanced systems and procedures to create high-quality products with minimum resource utilization also called the "flexible manufacturing operations," which is safe for the environment, users, and employees. Industry 4.0 monitors physical entities which are termed as "digital twins" of the process industries (Jamwal et al., 2021). In order to meet dynamic manufacturing goals and scenarios, economy should be more focused on the development of "smart" factories. In traditional manufacturing, conversion of raw material into final product is referred to as subtractive manufacturing, wherein raw materials are processed to final products by cutting, grinding, drilling, etc. Waste treatment technologies include bio-drying, gasification, dumping, landfarming, composting, and sewage treatment, which have a significant impact and can cause severe problems to the environment (Bànyai et al., 2019). It is essential to find solutions to achieve the objectives established by the United Nations Agenda for Sustainable Development 2030, and it is crucial to take the tenets of the circular economy into practice. A transformative approach offered by the circular economy is lowered emissions and a cleaner economy with restored biodiversity. A circular economy strategy along with 17 SDG goals are called the "heart" of all policymaking sectors (Fraga Lamas et al., 2021).

The connectivity that is needed for smart techs and smart computers on cloud platforms have been termed as "plug and play" in order to showcase the extent of high connectivity leading to reduction in time for processing of information. The circular economy promotes restorative design and the concept of "take make dispose recycle" design. World reports suggest that by 2050, the amount of waste generated could rise to 120 million tons. According to a global report, waste generation is projected to rise to 3.4 billion tons by 2050 (Shuai Mao et al., 2019). The increasing amount of waste generated is, therefore, calling for new techniques and innovations to cater to the urgent need. Systems like smart bins, waste sorting robots, waste monitoring, waste conveyor belts, waste streams, and smart grids to save energy and support resource recoveries are smart solutions to the emerging issues (James Clovis Kabugo et al., 2020). Therefore, AI is crucial for transitioning into a "zero circular waste economy" while taking into account the three main pillars of development, i.e., social, economic, and environmental (Fraga Lamas et al., 2021). Waste management is still a pressing issue that necessitates creative solutions for resource optimization, cost cutting, and, most importantly, sustainability. Figure 2.1 indicates that by improving collection, sorting, monitoring, and recycling procedures, AI is leading in improving waste management (Fang et al., 2023). Waste prevention, minimization, remediation, and reutilization are a few strategies that can be adopted for proper waste management. Recycling and reusing waste is an effective approach to minimize, reduce, and manage it. The discarded materials in landfills may be reused and remanufactured as well as repurposed. It, therefore, benefits the environment, conserves natural resources, saves energy, and decreases pollution. It acts in such a way that it ensures compatibility of the systems for seamless integration (David B. Olawade et al., 2024) and enhances as well as complements the abilities in order to increase resource recovery rate.

The primary aim of this chapter is to explore how smart technologies enhance the manufacturing sector's waste recycling process in terms of precision, sustainability, and efficiency along with how these technologies contribute to achieving objectives like energy efficiency, carbon footprint reduction, and zero waste. The contribution of this study is as follows:

- The study emphasizes on how human–machine collaboration helps in smart recycling plants.
- The research then analyzes how predictive analytics supports precision-driven waste management and resource efficiency. Following that, the study focuses on how AI and automation are transforming waste sorting and e-waste management.

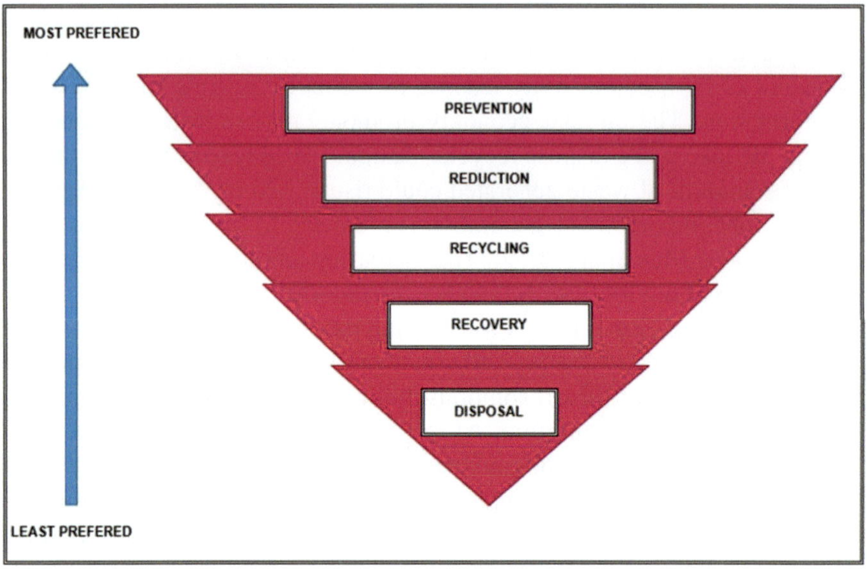

Figure 2.1 Least preferred method of disposal.

- The study deeply concentrates on smart grid integration: enhancing energy efficiency and hazardous waste generation.

2.2 Human–machine Collaboration in Smart Recycling Plants

The important components of Industry 4.0 are IoT, cloud computing, and cognitive computing. All these components collaborate to produce a smart environment. Industry 4.0 focuses on technology digitalization (Jena et al., 2019). Industry 4.0 has initiated a new era of cyber-physical systems (CPS) and interconnected smart manufacturing. The essential instructors of this framework are the technologies such as cloud computing, artificial intelligence, big data analytics, and the Internet of Things (IoT). By joining interconnected "things" such as sensors, actuators, controllers, robots, and machines, it becomes possible to manage system complexity while improving the performance of the entire production system (Alves et al., 2023).

In comparison with Industry 4.0, which prioritizes mass production, Industry 5.0 makes it easier to use resources and enables more efficient, customer-driven manufacturing solutions. Industry 5.0 is a method that focuses on people through three main pillars: human centricity, sustainability, and resilience. Industry 5.0 acknowledges the power of industry to attain societal goals beyond jobs and growth, becoming a resilient provider of prosperity – by making production respect the boundaries of the planet and by being a considerate contributor to sustainable development. Since the foundation of Industry 5.0 is more of a human-centric approach, it is gaining traction rapidly. Human–machine collaboration is given importance resulting in greater manufacturing (Xun Xu et al., 2021). Through the integration of innovative smart technologies with human knowledge, Industry 5.0 has made it feasible to develop sustainable and cost-effective manufacturing solutions.

Through human–machine collaboration, Industry 5.0 is an innovative approach that incorporates technologies including cloud manufacturing, intelligent healthcare, and the seamless incorporation as shown in Figure 2.2. By human–machine collaboration, researchers have examined how automation in warehouses and manufacturing could be profited more. The concept underlying Industry 5.0 is the idea to boost production effectiveness and versatility. On the other hand, Industry 4.0 emphasizes AI-driven technologies and digitalization as well as its core principles sustainability and equitable treatment. Industry 5.0 best describes the nature of the evolution of the industry as sociotechnical. Industry 5.0 is cross-sectoral and future-oriented which separates from the original configuration of Industry 4.0 (Arora et al, 2025; Ahma et al., 2025). Therefore, it is considered that Industry 5.0 is an evolution of Industry 4.0 "especially in the case of manufacturing and warehouse systems, human creativity and smart technologies will go hand-in-hand which will make the manufacturing industries rapid, efficient and more flexible." Since both industries (I4.0 & I5.0) remain extremely important, exploring their synergies can yield substantial insights into the continuous development of industry. Industry 5.0 promotes a more holistic purpose which is more concentrated toward human-centric approach and technology-driven paradigm. The essential distinction between Industry 4.0 and Industry 5.0 is the integration of humans and machines (Kumar et al., 2023). During the industrial era, the construction industry substantially contributed to the increase in carbon emissions in the atmosphere. Almost half of all energy and raw material use worldwide is attributed to construction. Thus, the construction industry is one of the sectors that significantly contributes to carbon emissions (Steffen Blömeke et al., 2020).

With the rising population of the world, the need for goods is also rising day-by-day. Researchers and the industries widely agree that future manufacturing should necessarily minimize its environmental impact. A researcher

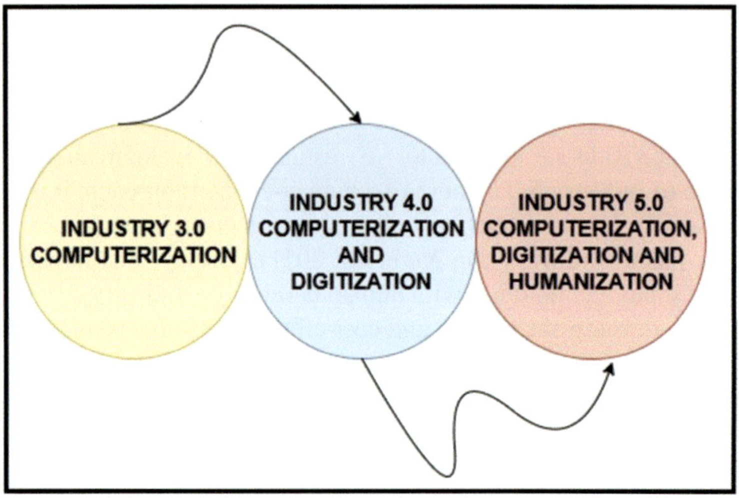

Figure 2.2 Transition from Industry 3.0 to 5.0.

defined this as the aim to "enhance the welfare of the human by safeguarding sources of raw materials which are required to meet human needs and verifying that the waste disposal stays within the sustainable limits, thereby avoiding harm to humans." Two essential paradigms that aim for enhanced environmental sustainability in the production process are Industry 4.0, also referred to as smart manufacturing and a circular economy (CE) (Kumar et al., 2024). By implementing automation to minimize manual errors, the above two paradigms aim to reduce production inefficiencies. An essential advantage of CE is the possibility to significantly reduce the environmental impact by replacing recycled or remanufactured components with newly created materials and products. This creates an important link between waste recycling procedures and smart manufacturing technology. Together, Industry 4.0 and CE indicate beneficial effects on sustainability (Pooja et al., 2024). While providing high-quality outcomes, remanufacturing's ability to reduce expenditure is of special importance. Therefore, remanufactured products with a lesser ecological footprint are likely to recover market share. Two effective methods for minimizing, declining, and managing trash are recycling and reusing. By recycling many of the leftover products that end up in landfills, the amount of waste sent to incinerators and dump sites could be drastically reduced while preserving natural resources. In manufacturing industries, waste recycling reduces the use of new raw materials and also lowers the environmental impact of waste treatment. Furthermore, during the

Table 2.1 Types of wastes and environmental impact.

S. No.	Types of waste	Explanation
1.	Waste glass	Glass waste has been recycled to make architectural mortar which can be more feasible, resilient, and has a greater capacity to compress. High-capacity lithium storage batteries have been constructed from it with the aim to preserve energy.
2.	Waste plastic	Plastic garbage is one of the major contributors to environmental pollution. The improper disposal and inadequate handling of plastic have severe long-term impacts on the environment, as it is a non-biodegradable material.
3.	Waste tire	As tires are made up of complex materials such as fiber, wire, and both natural and synthetic rubber, it is difficult to recycle them. Tires are also non-biodegradable, which causes negative impacts on the environment.
4.	Waste metal	For the purpose of generating new goods, metal recycling requires collecting, categorizing, cleaning, and melting scrap metal. The cost effectiveness, rising demand, growing conservation awareness, and environmental benefits of metal recycling have contributed to the recent increase in demand.
5.	Waste water	Waste water that comes from the commercial, industrial, and residential operations should be treated before releasing it into water bodies. Plants and animals will be harmed when waste is disposed of untreated.

creation of a product, recycling ensures that only a small amount of energy is used. Table 2.1 illustrated various types of waste and it environment impact.

2.3 Predictive Analytics: Precision-driven Waste Management and Resource Efficiency

This work aims to suggest an innovative Internet of Things-based and cloud-supported monitoring architecture for intelligent and smart manufacturing systems. It is designed to evaluate the overall system status and identify any irregularities within the production process. Additionally smart-driven technologies, including AI, detect specific anomalous events and offer corresponding risk classifications for potential solutions to address the event. A high degree of automation due to smart technologies is required for mass customization in industries, which thereby raises the cost of manufacturing and slows down production. To address this, advanced machine intelligence embedded in flexible manufacturing systems has been introduced in industries. This integration of smart tech sensors, along with Industry

4.0, aims to increase digital connectivity and resilience, along with a boost in operational efficiency and reduced downtime (Kathuria et al., 2024). AI plays a crucial role in improving waste management systems by analyzing data, identifying data, sorting as well as assisting in decision-making. AI is known for enhancing efficiency, boosting productivity, and aiding in boosting international environmental sustainability initiatives. In the waste recycling process, optimization occurs by minimizing expenses and maximizing material recovery to recycle and reuse as much as possible, preventing waste from going to landfills or incineration. Predictive analytics and real-time data monitoring are recent developments for streamlining the waste produced by the manufacturing industries. Many strategies that address design and operational challenges consist of statistical methodologies, simulations, simulation-based optimization, decision-making strategies, and integer programming.

Sustainability can be considered as the "core" of business strategies, as emphasized by The UN 2030 Agenda as shown in Figure 2.3. To help develop a balance between environmental responsibility and efficient manufacturing, smart manufacturing, energy efficiency, and low-impact industrialization are three major pillars of sustainable manufacturing. Production processes frequently generate scrap, which has a detrimental effect on both economic and environmental sustainability. Sensors are leveraged in assembly and sorting lines within manufacturing organizations for waste management, which helps improve economic sustainability and minimize emissions as well as waste within industries. Real-time production monitoring and optimization are boons of smart technologies in the manufacturing space, contributing to sustainability by increasing overall productivity, reducing waste, and optimizing resource use. Real-time data from process monitoring is essential for identifying potential deviations from the ordinary course of production and for optimizing processes accordingly. The dynamic propagation of these deviations is then analyzed by a reasoning engine, which further identifies various event streams and chains. The sensors that are embedded for predictive data analytics in process industries are defined as "invaluable tools" that provide insights into operation processes and measure of process variables for manufacturing a product that can be recycled for further use (Farooq et al., 2022). For example, a solid fuel's precise heating value numerical is essential to ensure seamless conversions and operations in waste-to-energy plants. This, in turn, prolongs the life of the manufacturing unit and enhances energy efficiency.

At first glance, zero waste refers to the complete elimination of waste, but it actually emphasizes reusing materials and preventing waste generation.

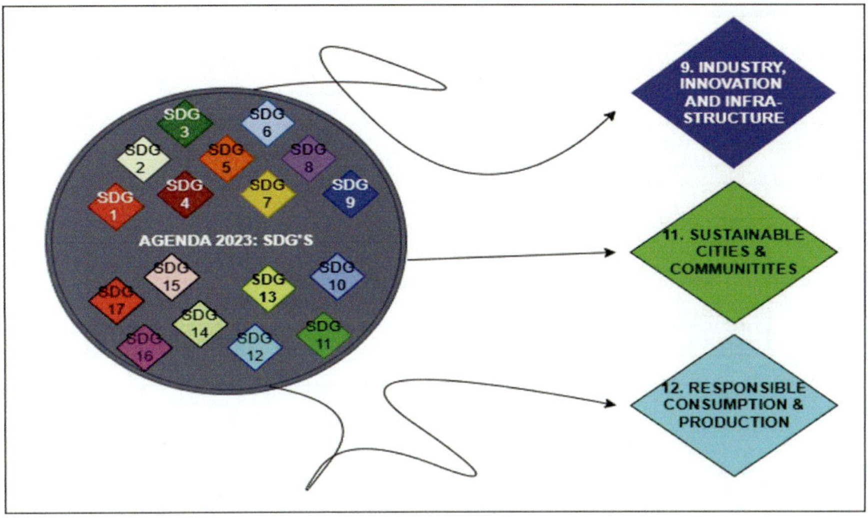

Figure 2.3 UN AGENDA 2030: Highlighting smart manufacturing.

This philosophy criticizes the idea of sending waste to landfills, dumpsites, or incinerators. An estimated 60 million barrels of oil-equivalent energy could be saved if all plastic content were removed from landfills. It lays down three principles revolving around responsible use, reuse, and recycling. This is supported by Zero Waste International Alliance (ZWIA) which is a global organization which encourages innovative solutions to preserve resources for environment friendly manufacturing methods as shown in Figure 2.4.

The deployment of Fourth Industrial Revolution technologies in the manufacturing sector has enhanced product functionality, reduced energy consumption, emissions of harmful gases, and, most importantly, increased productivity. Sustainable manufacturing involves a process of collection, sorting, breaking them down into smaller units, cleaning, and lastly selling them. Metals from most manufacturing units can be separated into ferrous and nonferrous streams, then melted accordingly in order to be used as other broken-down parts like tools, screws, and bolts. Smart recycling of metal leads to reduced operational costs and saves both energy and time through process optimization. While only 15% of the world's population lives in developed countries, they produce 34% of trash. Meanwhile, 23% of world's plastic comes from Asia, as shown in Figure 2.5.

A recent World Bank survey indicated that over 1.3 billion tons of waste are produced annually, and by 2025, it is predicted to increase to around

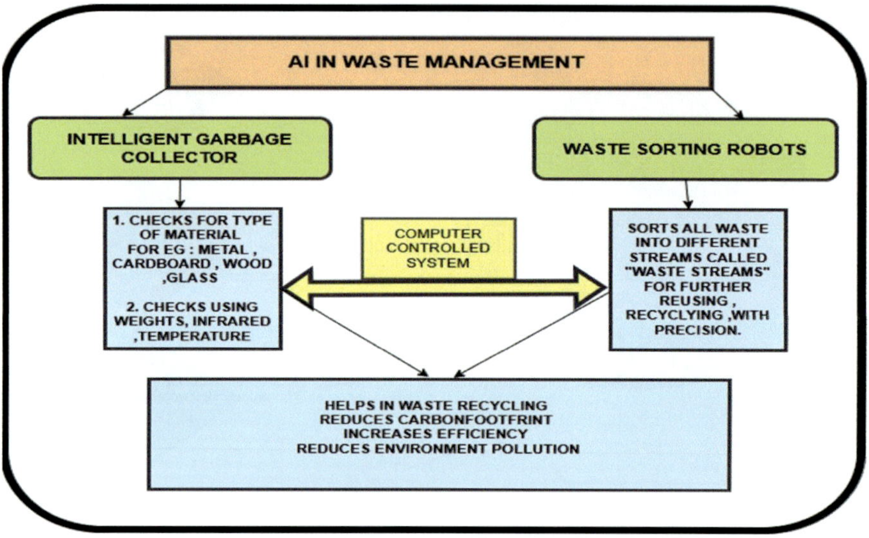

Figure 2.4 Process of AI in waste management.

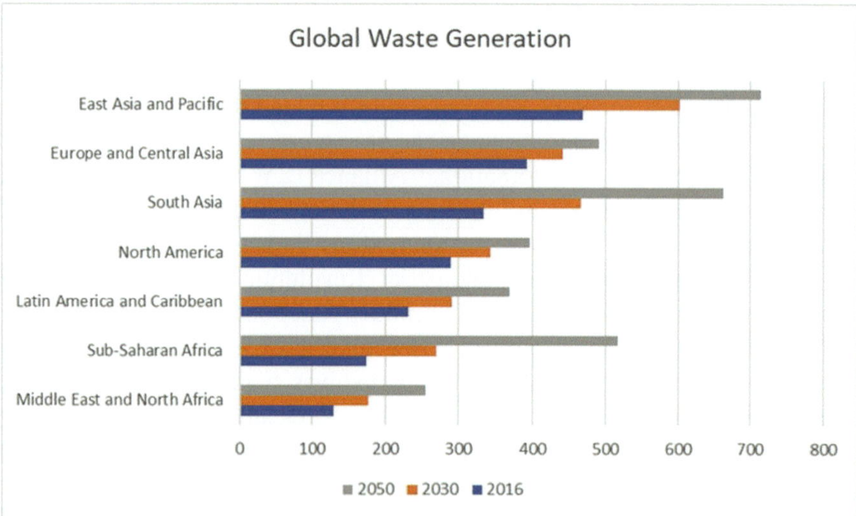

Figure 2.5 Global waste generation report.

2.2 billion tons. AI-based waste detection and inspection technologies help ensure composition and maintain the quality of recycled materials by detecting flaws, contamination, and noncompliance. Waste recycling facilities in industrial settings can schedule procedures to ensure effective resource usage,

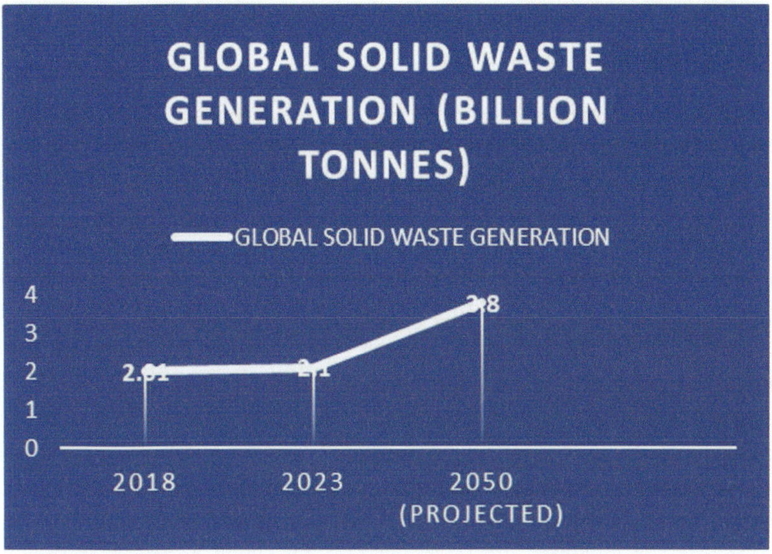

Figure 2.6 Global solid waste generation in billion tons.

reduce time waste generation, and optimize sorting, thanks to Ai's predictive capabilities. Additionally, the presence of contaminants in waste streams can interfere with the sorting processes, and using too many diverse materials can present a challenge (David B. Olawade et al., 2024). Smart bins are innovations which are widely used for urban purposes to regulate waste management by analyzing waste's weight, volume, and scheduling pickups for waste management authorities. Seas each year receive a total of 9.5 million tons of microplastics (main+microplastics) as shown in Figure 2.6 (Czekała et al., 2023). Manufacturing industries could take inspiration from the idea and use a larger version of it, which has a larger and smarter processor to classify different waste materials accordingly. This helps to contribute to their corporate social responsibility as well. Automated processes of waste sorting in the near future would help ensure higher precision, better categorization, and limit the need of manual labor, as well as minimize human error that could be generated. Waste from industries can be classified into solid waste, hazardous waste, liquid waste, and organic waste, each needing different treatment techniques. These waste types must be streamlined and managed with consideration for organized collection at the source, followed by optimized sorting. The use of smart technology and big data cloud computing programs to ensure maximum waste management is a need of the hour.

2.4 AI and Automation: Transforming Waste Sorting and E-Waste Management

Major concepts described by Industry 5.0 are the adherence to planetary boundaries and sustainability. The use of smart technology enables the industry to put forward a thorough overview of the entire manufacturing facility. These are based on three main pillars: "responsiveness," "connectivity," and "smartness." These pillars provide both an internal and external overview of the industry, facilitating smart cooperation and collaboration among manufacturing units while gathering data to effectively adapt to the dynamic nature of the industry. Industry 4.0 builds on fundamental aspects of industrial productivity and autonomy and has been defined as the "key" to Industry 5.0. Using smart technology combined with augmented human interaction and cloud computing facilitates operational flexibility, resilience, and a sustainable approach to industrialization, along with the seamless exchange of cross-organizational data. AI algorithms are thought to be the best models for accurate waste generation forecasts because of their data processing and predictive powers. AI sorting accuracy ranges from 72.8% to 99.95% (Mohammed et al., 2022).

To automatically sort and classify waste materials in order to meet recycling requirements, a digital system incorporating image processing, weight processing, temperature, density, and other different attributes is needed to effectively sort industrial waste in a way that can be reused as a consumption function as shown in Figure 2.7. Automating the industrial applications with smart technology solves object recognition problems, therefore since objects are recognised, it would be easy to sort them making waste management easier as it would help in the increase of recycled material. The main driving force behind this technology is the effort to reduce pollution, make a sustainable space, and create a societal awareness about the same. An automated smart technology helps sort waste from industries into five major categories: metal, wood, cardboard, glass, and others, managing it effectively through a model.

The system consists of several layers: physical layer, transmission layer, cloud cyber layer, real-time monitoring layer, and smart decision-making layer. Each layer manages a particular task on its own and processes information without altering the system as a whole. The physical layer involves manual tools for sorting, the transmission layer handles connectivity, the cloud cyber layer checks for any intrusions in the system before giving the results, the real-time monitoring layer monitors the waste and sorts it into

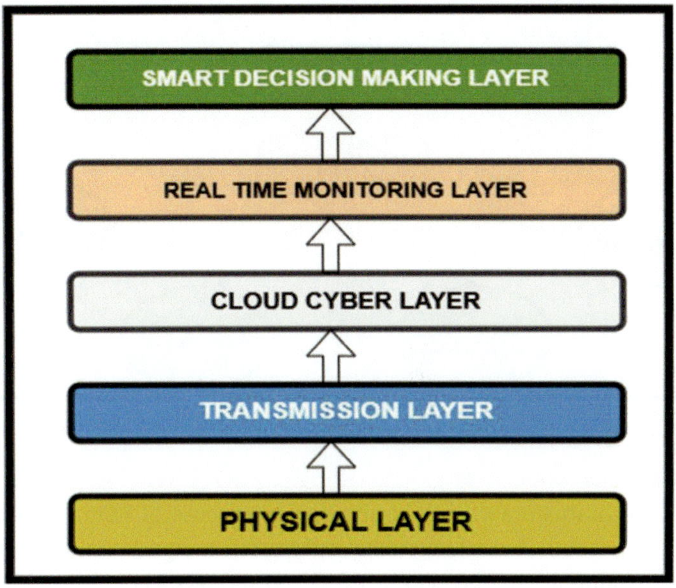

Figure 2.7 Five layers of smart machine.

streams, and finally, the smart layer provides information about the quality of the waste, its potential for recycling, and the method of recycling. This cloud-based, multi-layered system helps improve the accuracy and efficiency of recycling procedures. Crushed waste glass, when remanufactured, enhances durability and strength, and acts a strong substitute for fine aggregates. It follows the principle of "sustainable alternative" (Nautiyal et al., 2024).

Sorting waste into categories such as biodegradable, recyclable, and nonbiodegradable is necessary for efficient waste disposal. Agricultural productivity can be enhanced by repurposing biodegradable waste to enrich soil. Robots can physically interact with waste by recognizing visual characteristics like shape, color, and weight. This feature helps effective sorting which automatically directs materials to proper containers for additional recycling and is perfect for sorting wastes on conveyor belts. One popular kind of waste is e-waste, which needs disassembly, chemical treatment, and finally mechanical processing. For example, Bitcoin itself consumes almost as much power as a country. Source segregation is a method in which plastic, paper, glass, metal, and other wastes are separated at the very beginning of the process which leads to time and operational cost being saved. This could thereby help the current global recycling rate to go above 25% from 17.4% (Madhav

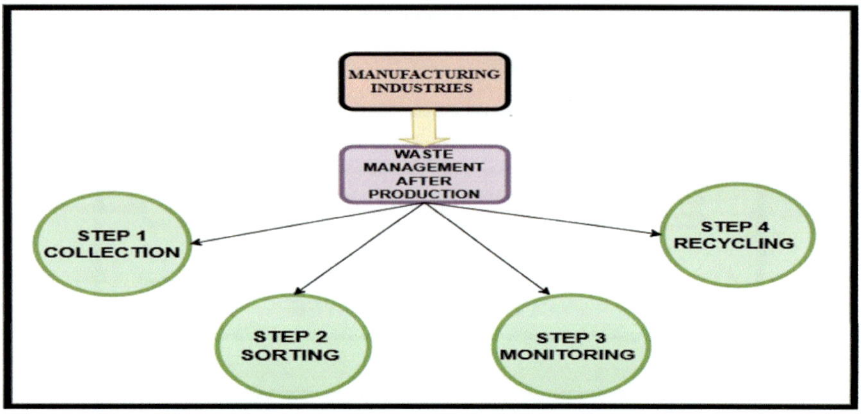

Figure 2.8 Steps in manufacturing industries for waste management after production.

et al., 2021). In this era of digital age, the global use of massive volumes of electronic devices has made e-waste generation a significant concern and a threat to sustainability. The output of electronic trash is expected to exceed 74.8 million tons by 2030, highlighting the urgent need for efficient recycling options. It is projected that e-waste will exceed 120 million tons by 2050.

Additional issues, such as water contamination from liquid waste in industries, continue to hinder recycling processes and lower the quality of recovered materials. The use of AI is trendy in one such matter which leads to firms arranging operating expenses by sorting and with the influence of smart bin systems to monitor waste levels (Negi et al., 2024). Precise identification and analysis of waste using technologies such as near-infrared spectroscopy and X-ray fluorescence help ensure proper disposal and recycling. AI-powered sorting systems support the circular economy by revolutionizing waste recycling through different layers and streamlining several phases of the process. Regular innovations in smart technologies provide advantages, ensuring compliance and more effective manufacturing systems. The entire steps that show how the process goes on for manufacturing industries for waste management after production are shown in Figure 2.8.

2.5 Smart Grid Integration: Enhancing Energy Efficiency and Hazardous Waste Generation

The smart grid has transformed the generation, transmission, distribution, and consumption of electricity, by integrating innovative sensing, communication, and control technologies into its daily operations. Cutting-edge advancements in power generation, delivery, consumption, and storage enhance effectiveness and reliability. Specifically, when it comes to the time-of-use (ToU) pricing models and the use of renewable energy sources, studies suggest that the smart grid could be used to improve energy efficiency (David et al., 2019). By merging the innovative technologies such as smart grid and smart metering, the emerging notion of smart cities puts a significant emphasis on smart energy and aims to directly strengthen local sustainable development. Energy-intensive organizations have to continuously enhance their energy efficiency. It has been recognized that minimizing energy consumption and maximizing energy efficiency could be accomplished by cleaner production. Energy-intensive manufacturing industries consume greater amounts of energy than other sectors as they rely upon extensive facilities and equipment. Smart manufacturing is an essential factor in increasing the quality of the process industry (Zhang et al., 2018). In smart manufacturing, incorporating innovative information technologies in inspection of process safety is becoming increasingly widespread. Smart manufacturing industries have transformed key areas:

1. Horizontal integration: strengthens the collaboration between the enterprises and agencies
2. Vertical integration: combining different organizational subsystems
3. End-to-end integration: integrating product design, customer necessities, and real-time alterations

By integrating all the above aspects, smart manufacturing enables mass modification, reduces costs, enhances sustainability, and increases production efficiency. Numerous technologies have been developed to address these issues, making green manufacturing an essential approach to ensuring process safety by reducing the use of energy and emissions. With the help of data-driven decision-making, intelligent early warning systems, smart monitoring, and optimization-based pollution reduction methods, green manufacturing analyzes every phase of the process lifecycle. While promoting long-term growth in the economy, it has the ability to substantially boost process industries' efficiency and safety (Kathuria et al., 2024).

Due to the widespread application of hazardous chemicals, green manufacturing is currently facing major issues related to safety management, resulting in spatial inhomogeneity in chemical processes and the need to comply with stricter safety and environment regulations. As a part of national plans resolving the new industrial revolution, green manufacturing is an essential component of smart manufacturing, which is now recognized as an important goal. In green manufacturing, efficiency and safety are given the highest importance. It highlights the necessity of laws regarding the environment and effective accident prevention techniques. The three primary challenges in green manufacturing are as follows:

1. Integration of information
2. Support for decision-making and dynamic risk assessment
3. Early warning systems

The two important enabling technologies in Industry 4.0 are digital twins and analytics. An organization's social goals and economic sustainability will strengthen by the improvements in the Industry 4.0. These technologies offer tremendous potential to foster sustainable value creation in the social, environmental, and economic domains by optimizing the efficiency of resources. The researchers have examined the ways Industry 4.0 influences sustainable manufacturing. However, they have reached the conclusion that these technologies will eventually open up new business opportunities for sustainability.

A new era of sustainability, digital transformation, and intelligence is arising in modern manufacturing which results in the rapid growth of sensing, communication, computing, and analytics technologies. For the sustainable development of smart cities, smart energy and technology are important. In the current industry, sustainable manufacturing and smart manufacturing are the two significant subjects that have a lot of similarity. In the light of the present-day resource consumption and environmental issues, the main aim of smart manufacturing is sustainability. Enhancing energy efficiency is one of the primary objectives of smart manufacturing as it leads to higher energy savings and minimizes resource consumption, making it an effective way for sustainability. The basic principles of sustainable manufacturing are as follows:

1. Minimal or barely any harm to the society and environment
2. Preservation of natural resources
3. Capability to satisfy present and future energy needs
4. High operational efficiency
5. No hazardous substances are discharged into air, land, or water
6. Minimal or negligible emissions of greenhouse gases
7. No additional pressure on future generations

The source-system-service (3S) has been suggested in smart manufacturing to achieve sustainability. The "source" stage of this approach provides an enormous value on utilizing resources and energy which are abundant, clean, renewable, affordable, and ecologically friendly. The European Union has allocated $7.8 billion under its "Factories of the Future" program to promote both smart and sustainable manufacturing. China has established green manufacturing as an essential national endeavor. The objective of this initiative is to develop manufacturing processes that are cleaner and ecologically friendly (Hoang et al., 2021).

2.6 Discussion and Suggestions

In many countries, waste recycling and resource recovery are not regulated properly and the political interests have periodically influenced in what ways these laws should be implemented. To safeguard the environment, policies prioritizing full resource recovery, 100% waste recycling, and the complete elimination of landfills and incineration should be given the utmost importance. The use of innovative and environmentally friendly technologies helps enhance tracking and documentation processes, reduce the risk of contamination, and provide cost-effective solutions. Even so, certain waste techniques and technologies could be challenging, laborious, and often more expensive than forming new materials. Artificial intelligence (AI) is being integrated into sustainable waste management to substantially reduce natural resource usage without compromising quality of life. During the disposal procedure, inefficient waste disposal practices result in considerable pollution of the environment, excessive expenditure, and lack of oversight. Waste management is still an important concern for countries that are industrialized as well as developing. On the contrary, AI has the capacity to further enhance efficiency of waste treatment, decrease environmental damage, and give more sophisticated computational solutions for more intelligent and effective waste management processes. Usage of smart technologies has become essentially important as there is not much room left on earth for disposal of waste. Before shifting to recycling, the 5Rs framework – refuse, reduce, reuse, repurpose, and recycle – serves as a mentor to reduce waste. For the ethical use of AI, clear regulations should be developed. These frameworks ought to tackle the challenges like security, fairness, honesty, and data privacy. System interoperability promotes AI adoption and sharing of the data could be aided by the policy measures.

In the integration of Industry 4.0, i.e., smart factories, the opportunities for boosting efficiency and reliability of remanufacturing procedures are provided. Furthermore, while improvising the quality of remanufactured products, smart technologies help to minimize the cost. The limitations are described briefly below:

1. A single classifier was applied for evaluating automated waste sorting and recycling classification by using an artificial neural network (ANN) and feature fusion. For the purpose of providing a more in-depth assessment, it is essential to test more classifiers (Camarinha-Matos et al., 2022).
2. The presented automated machine for recycling and waste sorting depends upon the feature fusion which has limitations such as high computational cost and time complexity.
3. In order to include other waste categories such as COVID-19-related waste and other medical waste, the classification model has to be upgraded.

In order to enhance the accuracy of ANN-based methods, tactics such as fine-turning and data augmentation can be used in future studies. Moreover, the ANN's performance could be boosted by extending the quantity of training data. In general, this study demonstrates how waste management procedures can be enhanced by digitally driven a circular economy projects. The numerous strategic agendas across the diverse locations have outlined the issues of sustainable and resilient manufacturing. Guided by the sustainable development goals of the UN 2030 Agenda, a number of these agendas are geopolitically driven and often serve as policy suggestions rather than wholly research-focused frameworks. Yet, they continue to be important in directing conversations and bringing the concept together in a way to create a vision for resilient and sustainable manufacturing systems. Barriers such as waste heterogeneity, surface contamination, uneven masses and shapes, and the random positioning of items in the waste stream make it challenging to implement robotic technologies for waste sorting. Under these circumstances, errors may occur when the actuator misinterprets the object's location and fails to grasp it properly. Guidelines could also be helpful in establishing proper waste management processes (Sarc et al., 2019).

Tata steel, one of the largest and top steel manufacturers in India has put in place recycling programs. It consists of recycling scrap metal and reducing its primary iron ore. Green manufacturing practices reduce negative effects on the environment and minimize waste generation (Chemco Group, 2024).

Since 2008 almost 75% of the garbage produced by Intel's activities has been recycled. By 2020, Intel aims to divert all hazardous trash from landfills and recycle 90% of it (Intel Corporation, 2023-24). All sites, including manufacturing, distribution, and innovation departments, have achieved zero industrial waste to landfill (Just Capital, 2018), (GlobalData Estée Lauder, 2021). Uses organic cotton and materials like fish nets to make clothes. Leader in sustainable fashion. Strong emphasis on environmental responsibility. Runs Denim recycling programs, encourages customers to recycle used jeans, also prioritizes water conservation in manufacturing processes (Climate Seed Sustainable Fashion, 2022). Founded in 1989, focuses on medical waste management, recycling initiates, protects public health through innovative waste solutions, and properly dispose of regulated trash for healthcare facilities (Climate Seed Sustainable Fashion, 2022).

2.7 Conclusion

Due to the increase in pricing and adverse environmental impact, waste minimization, reduction, conversion, and recycling techniques have gained a lot more attention. The chapter has thoroughly examined the methods and technologies that could be used to attain zero waste management as well as the contributions of innovative Fourth Industrial Revolution (I4.0) and Fifth Industrial Revolution (I5.0) technologies in manufacturing.

AI-powered sorting and monitoring have revolutionized recycling processes by improving productivity and efficiency along with lowering contamination in the waste stream. Cloud-based waste management helps in processing large amounts of data, facilitates data sharing and automation which makes it easier to recycle waste effectively. E-waste is increasing at an alarming rate due to people's tendency to buy new products; therefore, looking into this growing issue is the need of the hour. Maximizing resource recovery and reducing landfill trash are some advantages of incorporating AI into manufacturing systems making the industry "smart."

The amalgamation of human and AI helps in translating complex waste streams into streams that can be easily sorted, therefore reducing the carbon footprint, as most of the waste is recycled with precision sustainably. These are goals that support the goals of a circular economy. It is important to minimize energy usage in recycling and waste treatment facilities to contribute to social responsibility. Industries can regenerate energy using different techniques, which therefore lowers carbon emissions and operational costs. Therefore, the integration of smart and green manufacturing, along with keeping in mind the UN Agenda 2030, is the ultimate goal to be more resource efficient as well as energy efficient.

Human-centred automation portrays a holistic approach to mark the shift from Industry 4.0 to Industry 5.0, which is developing in a way that is social and equally environmental and economical.

References

Caiazzo, B., Murino, T., Petrillo, A., Piccirillo, G., & Santini, S. (2022). An IoT-based and cloud-assisted AI-driven monitoring platform for smart manufacturing: design architecture and experimental validation. Journal of Manufacturing Technology Management, 34(4), 507–534.

João Barata, Ina Kayser, Industry 5.0 – Past, Present, and Near Future, Procedia Computer Science, Volume 219, 2023, Pages 778-788, ISSN 1877-0509.

Jamwal, A., Agrawal, R., Sharma, M., & Giallanza, A. (2021). Industry 4.0 Technologies for Manufacturing Sustainability: A Systematic Review and Future Research Directions. Applied Sciences, 11(12) , 5725.

Bányai, T., Tamás, P., Illés, B., Stankevičiūtė, Ž., & Bányai, Á. (2019). Optimization of municipal waste collection routing: Impact of Industry 4.0 technologies on environmental awareness and sustainability. International Journal of Environmental Research and Public Health, 16(4), 634.

Fraga-Lamas, P., Lopes, S. I., & Fernández-Caramés, T. M. (2021). Green IoT and Edge AI as Key Technological Enablers for a Sustainable Digital Transition towards a Smart Circular Economy: An Industry 5.0 Use Case. Sensors, 21(17), 5745.

Shuai Mao, Bing Wang, Yang Tang, Feng Qian, Opportunities and Challenges of Artificial Intelligence for Green Manufacturing in the Process Industry, Engineering, Volume 5, Issue 6, 2019, Pages 995-1002, ISSN 2095-8099.

James Clovis Kabugo, Sirkka-Liisa Jämsä-Jounela, Robert Schiemann, Christian Binder, Industry 4.0 based process data analytics platform: A waste-to-energy plant case study, International Journal of Electrical Power & Energy Systems, Volume 115, 2020, 105508, ISSN 0142-0615.

Fraga-Lamas, P., Lopes, S. I., & Fernández-Caramés, T. M. (2021). Green IoT and Edge AI as Key Technological Enablers for a Sustainable Digital Transition towards a Smart Circular Economy: An Industry 5.0 Use Case. Sensors, 21(17), 5745.

Fang, B., Yu, J., Chen, Z., Osman, A. I., Farghali, M., Ihara, I., Hamza, E. H., Rooney, D. W., & Yap, P. (2023). Artificial intelligence for waste management in smart cities: a review. Environmental Chemistry Letters, 21(4), 1959–1989.

David B. Olawade, Oluwaseun Fapohunda, Ojima Z. Wada, Sunday O. Usman, Abimbola O. Ige, Olawale Ajisafe, Bankole I. Oladapo, Smart waste management: A paradigm shift enabled by artificial intelligence, Waste Management Bulletin, Volume 2, Issue 2, 2024, Pages 244-263, ISSN 2949-7507.

Jena, M. C., Mishra, S. K., & Moharana, H. S. (2019). Application of Industry 4.0 to enhance sustainable manufacturing. Environmental Progress & Sustainable Energy, 39(1).

Alves, J., Lima, T. M., & Gaspar, P. D. (2023). Is Industry 5.0 a Human-Centred Approach? A systematic review. Processes, 11(1), 193.

Xun Xu, Yuqian Lu, Birgit Vogel-Heuser, Lihui Wang, Industry 4.0 and Industry 5.0—Inception, conception and perception, Journal of Manufacturing Systems, Volume 61, 2021, Pages 530-535, ISSN 0278-6125.

Steffen Blömeke, Julian Rickert, Mark Mennenga, Sebastian Thiede, Thomas S. Spengler, Christoph Herrmann, Recycling 4.0 – Mapping smart manufacturing solutions to remanufacturing and recycling operations, Procedia CIRP, Volume 90, 2020, Pages 600-605, ISSN 2212-8271.

Kathuria, S., Duggal, A. S., & Lande, J. (2024, May). Artificial Intelligence Automation in Advent of Human Resource Management. In *2024 Parul International Conference on Engineering and Technology (PICET)* (pp. 1-5). IEEE.

Farooq, M., Cheng, J., Khan, N. U., Saufi, R. A., Kanwal, N., & Bazkiaei, H. A. (2022). Sustainable Waste Management Companies with Innovative Smart Solutions: A Systematic Review and Conceptual Model. Sustainability, 14(20), 13146.

Czekała, W., Drozdowski, J., & Łabiak, P. (2023). Modern Technologies for Waste Management: a review. Applied Sciences, 13(15), 8847.

Mohammed, M. A., Abdulhasan, M. J., Kumar, N. M., Abdulkareem, K. H., Mostafa, S. A., Maashi, M. S., Khalid, L. S., Abdulaali, H. S., & Chopra, S. S. (2022). Automated waste-sorting and recycling classification using artificial neural network and features fusion: a digital-enabled the circular economy vision for smart cities. Multimedia Tools and Applications, 82(25), 39617–39632.

Nautiyal, R., Jha, R. S., Kathuria, S., Pandey, S., Singh, R., Gehlot, A., ... & Alkhayyat, A. (2024). Artificial intelligence's foresight in cybersecurity.

In *Artificial Intelligence and Blockchain in Industry 4.0* (pp. 38-51). CRC Press.

Madhav, A. S., Rajaraman, R., Harini, S., & Kiliroor, C. C. (2021). Application of artificial intelligence to enhance collection of E-waste: A potential solution for household WEEE collection and segregation in India. Waste Management & Research the Journal for a Sustainable Circular Economy, 40(7), 1047–1053.

Negi, P., Singh, R., Gehlot, A., Kathuria, S., Thakur, A. K., Gupta, L. R., & Abbas, M. (2024). Specific soft computing strategies for the digitalization of infrastructure and its sustainability: A comprehensive analysis. *Archives of Computational Methods in Engineering*, *31*(3), 1341-1362.

David, M., & Koch, F. (2019). "Smart is not smart enough!" Anticipating critical raw material use in smart city concepts: the example of smart grids. Sustainability, 11(16), 4422.

Zhang, Y., Ma, S., Yang, H., Lv, J., & Liu, Y. (2018). A big data driven analytical framework for energy-intensive manufacturing industries. Journal of Cleaner Production, 197, 57–72.

Kathuria, S., Rawat, P., Kumar, G. R., Singh, B., & Taneja, A. (2024, May). Amalgamation of Technologies for Smart Monitoring of Sustainable Aquaculture. In *2023 International Conference on Smart Devices (ICSD)* (pp. 1-5). IEEE.

Hoang, A. T., Pham, V. V., & Nguyen, X. P. (2021). Integrating renewable sources into energy system for smart city as a sagacious strategy towards clean and sustainable process. Journal of Cleaner Production, 305, 127161.

Camarinha-Matos, L. M., Rocha, A. D., & Graça, P. (2022). Collaborative approaches in sustainable and resilient manufacturing. Journal of Intelligent Manufacturing, 35(2), 499–519.

Sarc, R., Curtis, A., Kandlbauer, L., Khodier, K., Lorber, K., & Pomberger, R. (2019). Digitalisation and intelligent robotics in value chain of a circular economy oriented waste management – A review. Waste Management, 95, 476–492.

Chemco Group. (n.d.). Case studies of successful recycling initiatives in India. Chemco Group. https://blog.chemcogroup.com/news/case-studies-of-successful-recycling-initiatives-in-india/ (Accessed on 2nd March, 2025)

Intel Corporation. (n.d.). Corporate responsibility. Intel. Available on: https://www.intel.com/content/www/us/en/corporate-responsibility/corporate-responsibility.html (Accessed on 5th March, 2025)

Just Capital. (2018, April 20). These 5 companies are leading the charge on recycling. Forbes. Available on: https://www.forbes.com/sites/justcapital/2018/04/20/these-5-companies-are-leading-the-charge-on-recycling/ (Accessed on 5th March, 2025)

GlobalData. (n.d.). Estée Lauder waste generation. Global Data. Available on: https://www.globaldata.com/data-insights/consumer/estee-lauder-waste-generation-2095822/#:~:text=The%20total%20waste%20generated%20by,by%20the%20company%20was%20hazardous (Accessed on 5th March, 2025).

Arora, M., Ahmad, V., Kumar, R., & Singh, R. (2025). The Impact of Industry 4.0 Technologies on Operational Excellence Methods: Application of Fuzzy Topsis. Engineering Management Journal, 1-12.

Ahmad, V., Arora, M., Sayal, A., & Kumar, R. Exploring Trends in E-Business and Sustainable Finance: An Integrated Analytical Approach. Frontiers in Sustainable Cities, 7, 1549874.

Devi, P., Desai, S. S., Kumar, R., Naruka, M. S., & Tripathi, V. (2024, December). A Strategic Innovations for Stock Market Optimization by Using Data Science & Explainable AI in High Frequency Trading. In 2024 International Conference on IoT Based Control Networks and Intelligent Systems (ICICNIS) (pp. 1510-1515). IEEE.

Kumar, R., Goel, R., Singh, T., Mohanty, S. M., Gupta, D., Alkhayyat, A., & Khanna, R. (2024). Sustainable Finance Factors in Indian Economy: Analysis on Policy of Climate Change and Energy Sector. Fluctuation & Noise Letters, 23(2).

Kumar, R., Kandpal, B., & Ahmad, V. (2023). Industrial IoT (IIOT): Security threats and countermeasures. 2023 International Conference on Innovative Data Communication Technologies and Application (ICIDCA), 829–833. IEEE.

Climate Seed. (n.d.). Sustainable fashion: How to reduce your environmental impact. Climate Seed. Available on: https://climateseed.com/blog/sustainable-fashion-how-to-reduce-your-environmental-impact. (Accessed on 5th March, 2025).

Angel One. (n.d.). Best waste management and recycling companies. Angel One. Available on:https://www.angelone.in/news/best-waste-management-and-recycling-companies. (Accessed on 7th March, 2025)

3

Artificial Intelligence in Waste Management: Optimizing Recycling, Cost Reduction, and Sustainability

Girija Nandini[1], Swetalina Mishra[1], Bijay Kumar Paikaray[2], and V. S. Damodharan[3]

[1]School of Management, Centurion University of Technology and Management, India
[2]Centre for Data Science, Department of Computer Science and Engineering, Siksha 'O' Anusandhan (Deemed to be) University, India
[3]Department Chair Business, Abu Dhabi Vocational Education & Training Institute; United Arab Emirates
E-mail: irijanandini@cutm.ac.in; Psweta07in@gmail.com, bijaypaikaray87@gmail.com; sriramdams@gmail.com

Abstract

Waste management is a global contest that requires innovative technology for sustainability and a circular economy. Traditional models are not sufficient to solve the problem of waste management. In this scenario, new technologies like artificial intelligence (AI) can solve this complex problem in manufacturing companies. This research will analyze the role of AI in waste management, including assembly, arrangement, reprocessing, and controlling. It demarcates the possible advantages and problems of using AI and highlights the enhanced quality of information, confidentiality, reduction of cost, and moral deliberations. AI holds a substantial role in effective waste management by solving issues like confidentiality concerns, data quality, and reduction of cost. Optimum utilization of resources and sustainability can be achieved through AI-driven waste management systems. This research connects the information gap and simplifies the understanding of the use of artificial intelligence for waste management. It will support budding research,

DOI: 10.1201/9788743806905-3

policymakers, governments, municipalities, and also organizations dealing with waste management.

Keywords: Machine learning, waste management, sustainability, artificial intelligence.

3.1 Introduction

In the current scenario, advanced and effective solutions are highly essential for achieving sustainable goals through responsible waste management. However, there are no standardized global waste management practices, creating problems in efficiency (Olawade et al., 2024). Additionally, due to the complication of waste composition, harmful materials make recycling and dumping problematic (Wang et al., 2022). Different types of waste are creating problems for the environment by affecting soil, air, and water. Wastes are mainly created because of production companies, agricultural production pollution releases, and consumption (Chen et al., 2023; Devi Pooja et al., 2024). It will create ecological pollution, financial losses, health problems, and increased cost. To solve this problem, the government, policymakers, and different companies started sustainable waste management practices. Furthermore, awareness about sustainable waste management will reduce waste generation and individuals can make better decisions to achieve the sustainable goal (Osman et al., 2023). The classification of waste is different and categorized based on its physical form, state, or composition (Peng et al., 2023). Industrial waste is a major source of waste that creates problems in the environment. This waste can be wastewater, scraps produced after the production, or slag containing radioactive ingredients (Patel et al., 2022; Tawfk et al., 2022). Moreover, agricultural waste is also a major part of total waste creating pollution for the environment. This waste can be reused for composting, for adding value, or can be used for production after recycling. The waste can be solid waste or waste which are dangerous to society (Jha et al., 2022; Peng et al., 2023). Moreover, insufficient public awareness regarding appropriate waste disposal creates problems in the health of the community and proper waste management (Ramya et al., 2024). By using AI, waste management processes, decision-making, sustainability, and resource optimization can be improved (Hu et al., 2022). Nowadays, the use of AI and its impact is growing for the management of waste in different regions (Pallathadka et al., 2023). AI is transforming waste management systems with advanced technologies, future analysis, and smart waste collection methods (Krishna

& Sharma, 2023; Wilts et al., 2021). Traditional waste management methods are inefficient and irregular (Cubillos, 2020). Nevertheless, by using AI these problems can be solved by optimizing the waste collection ways (Yadav & Karmakar, 2020; Olawade et al., 2024). AI can monitor real-time waste levels with the smart bin system for efficient waste management (Hussain et al., 2020). Traditional systems are manual and lead to more errors, time, and problems in the management system (Lin et al., 2022; Guo et at., 2021). AI techniques like machine vision and image recognition for waste storage increased its accuracy making the waste management system more efficient (Zhang et al., 2021). Moreover, a robotic sorting system can sort the waste material effectively making the recycling process better. This process will maximize the resource allocation and reduce the cost (Wu et al., 2023). This study will examine the role of AI in waste management, including assembly, arrangement, reprocessing, and controlling. It demarcates the possible advantages and problems of using AI and highlights the enhanced quality of information, confidentiality, reduction of cost, and moral deliberations. This research connects the information gap and simplifies the understanding of the use of artificial intelligence for waste management.

3.2 Methodology

This chapter integrates several fundamentals to bring a widespread understanding of the use of AI for the management of waste. The study is focusing on literature study, utilizing peer-reviewed journals, industry reports, and conference proceedings. This review evaluates recent studies on AI-driven waste management. Mostly, the publications from 2021 to 2023 are taken from Google Scholar and Scopus databases to include pertinent literature. This review delves into definite areas like waste collection, waste sorting, and waste monitoring which are important for waste management practices.

3.3 AI Techniques Used for Waste Management

AI techniques are highly useful for managing large amounts of waste and giving better results (See Figure 3.1). Different AI techniques are used for effective sustainable waste management. Artificial neural networks (ANNs) copy and behave like a human brain, used for optimizing waste generation, waste sorting, and overall waste management systems (Jassim et al., 2023). Linear regression (LR) is mostly used for future prediction of solid waste management (SWM) and for optimizing waste generation

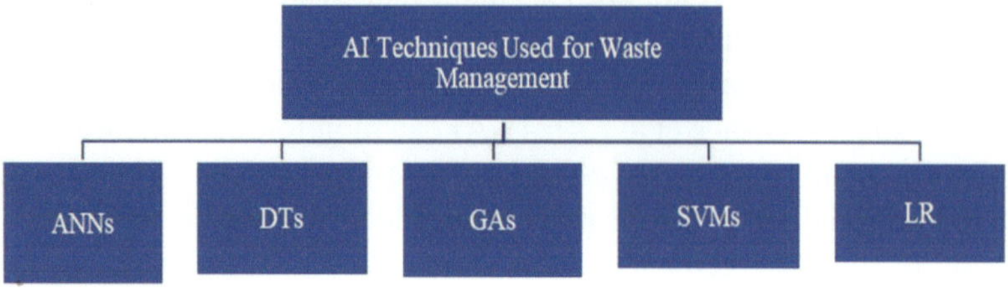

Figure 3.1 Use of AI techniques for waste management.

(Huang et al., 2023). Support vector machines (SVMs) are a technique used in waste generation to increase efficiency (Altin et al., 2023). Decision trees (DTs) are a machine learning technique that is highly efficient for predicting large datasets used for the prediction of waste generation and optimization (Massoud et al., 2023).

A genetic algorithm (GA) is an optimization technique used to plan waste management (Thaseen et al., 2023). Additionally, AI can predict the future forms of waste generation, supportings decision-making and resource apportionment to manage the growing volume of waste. AI will give solutions and adaptable methods to manage global waste problems (Ahmad Vasim et al., 2025).

3.3.1 AI for waste collection

AI is used in waste collection to enhance efficiency, thereby confirming and supporting sustainability. Different models are used to make this process more efficient and effective (Sharma & Vaid, 2021) (see Figure 3.2). IoT technology with smart bin systems collects real-time data and also uses machine learning algorithms for future prediction and optimization of waste collection (Bijos et al., 2022). Waste collection agendas constructed on real-time figures and fluctuating situations use active scheduling for better results. Waste management companies are collecting data by using AI technologies and optimizing the collection schedules, efficiently distributing the resources. Predictive modeling and analytics are utilized to detect forms, movements, and associations in the composed information (Solano Meza et al., 2023). Forecasting models use machine learning and collect data from social media, IoT sensors, and online platforms for future prediction. These models consider different factors like past waste generation rates, population density of a particular place, events, holidays, and demographic data to forecast waste generation for a future period. Furthermore, by forestalling variations

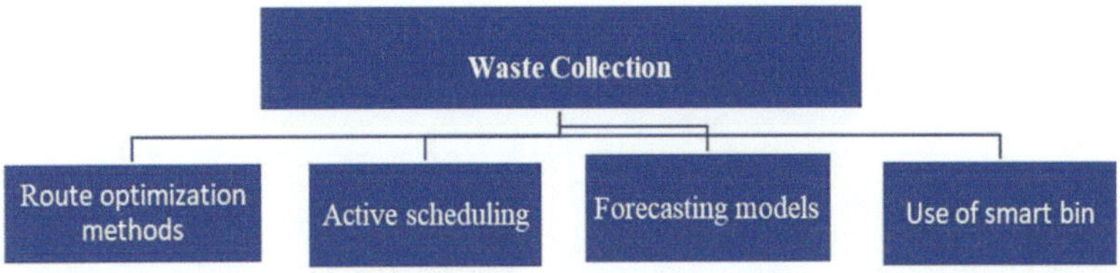

Figure 3.2 Use of AI for waste collection.

in waste generation, authorities can make faster decisions and distribute resources properly. It needs cautious scheduling and deliberation to optimize its efficiency and diminish possible problems (Andeobu et al., 2022). By using route optimization methods, travel time, vehicle emissions, and fuel consumption can be reduced during waste collection. Route optimization is possible with the use of machine learning by getting real-time data about traffic situations, weather forecasts, and historical collection of data (Munir et al., 2023).

3.3.2 AI for waste sorting

Waste sorting is required for proper recycling, where waste is seperated according to its categories. AI-driven waste, sorting process has revolutionized the waste management system. Automated sorting technologies use advanced sensors to classify and sort waste materials according to their composition (Yan et al., 2021). Image recognition and machine vision require high-resolution cameras for analysis and then separating the waste materials according to their composition to make the recycling process easier and more accurate (Baduge et al., 2022). Robotic sorting systems and sensor techniques separate waste according to their composition and pick leftover substances according to their composition (Subramanian et al., 2021; Feng et al., 2022) (see Figure 3.3).

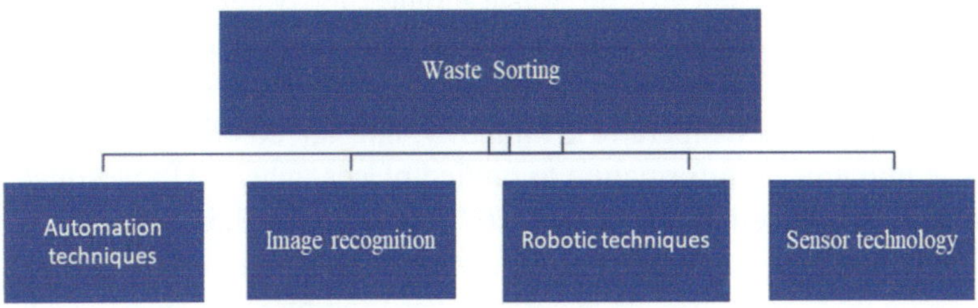

Figure 3.3 Use of AI for waste sorting,

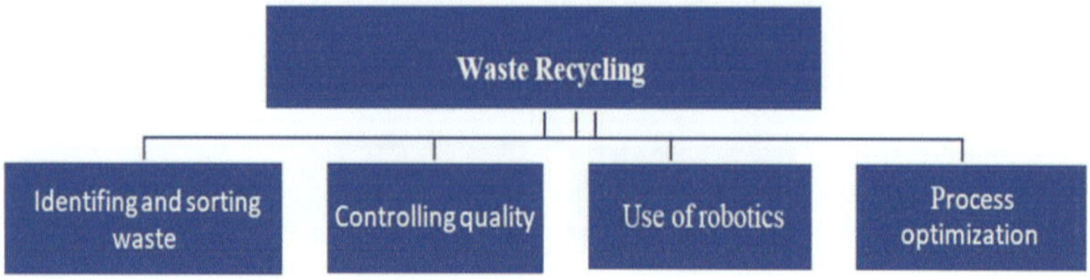

Figure 3.4 Use of AI for waste recycling.

3.3.3 AI for waste recycling

AI technologies have taken a substantial role in refining several features of recycling practices. AI uses advanced technologies to identify and sort complex waste and mixed materials based on their composition. It increases the competence and transparency of recycled material (Chauhan et al., 2023). AI-based quality control and inspection systems use machine learning methods, and sensors, to remove impurities from recycled materials and also from the environment. It will increase the marketability and reuse of recycled materials. The AI-driven technology will classify the material according to their quality leading to enhanced efficiency (Pluskal et al., 2021). The usage of robotics modernizes processes, upsurges quantity, decreases employee costs, and advances total resource retrieval rates (Mao et al., 2021). Additionally, process optimization for waste recycling involves AI, machine learning models, and predictive analysis that monitor real-time data, leading to reduced cost and improved sustainability in recycling operations (Chidepatil et al., 2020) (see Figure 3.4).

3.3.4 AI for waste monitoring

With the adoption of AI technology, waste monitoring has experienced substantial revolutions. Real-time monitoring systems utilize sensors for collecting data, collection vehicle routes, and the disposal place. It will increase the operational efficiency of the waste monitoring system (Salman & Hasar, 2023). It helps the authorities in policy decision-making in the infrastructure (Ali et al., 2021). IoT and sensors collect data regarding several factors, like the level of waste, temperature, humidity, and air condition. These instruments communicate information to unified places, where AI investigates and communicates information (Bernat, 2023) (see Figure 3.5).

Predictive analysis predicts future waste generation trends and patterns. This technology increases its accuracy over time by analyzing the new data continuously (Reza, 2023).

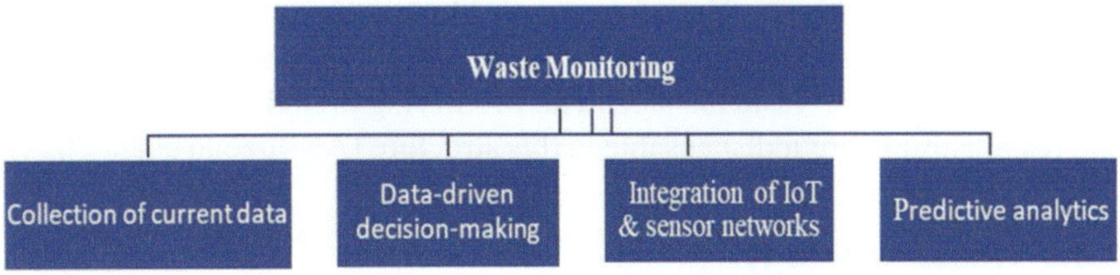

Figure 3.5 Use of AI for waste monitoring.

3.3.5 Problems of AI-driven waste management

Though AI has a major role in sustainable waste management practices, it has its problems to be addressed for effective waste management (see Figure 3.6).

AI-driven waste management technology highly depends on accurate real-time data. The collection of accurate data is a major challenge for AI-driven waste management. A collaborative initiative between the stakeholders is required to solve the issue (Tsui et al., 2023). Data secrecy and safety are also a major concern for AI-driven waste management systems. A balance should be maintained between the use of AI and the privacy and security of sensitive information (Edinov and Fauzi, 2023). Encryption techniques should be adopted to guard sensitive information while storing it. Furthermore, applying admission controls and user verification machines can confine illegal admission to information. Consistent safety audits and susceptibility valuations are indispensable to recognize and mitigate possible safety

Figure 3.6 Problems of AI-driven waste management

intimidations proactively. The adoption of AI technology requires high cost with supported infrastructure. Moral considerations are one of the important aspects of AI-driven waste management. Modifying prejudices, transparency, and responsibility are indispensable in this structure (Al-Ruzouq et al., 2022).

3.3.6 Waste to energy

Electricity can be generated from biogas using organic waste produced by animals and humans, leading to sustainability. The consumption of energy will rise in the future, making the creation of renewable energy essential (Farghali et al., 2023). Economic development and a circular economy can be possible with the creation and use of renewable energy. Among all renewable energy sources, biogas is one of the most important, r as it is environmentally friendly and sustainable over the long term (Chiu et al., 2022). Though the production of biogas requires complex inputs and measured response times, it is highly important for the circular economy. In the current situation, researchers are using predictive models to identify the important factors for biogas production. Also, machine learning techniques are used to increase the output of biogas production and reduce problems encountered during the producing process (Pence et al., 2023). In the present scenario, machine learning, deep learning, and artificial neural networks are combined to achieve better results and output (Kumar. R et al., 2024). It can solve the problems related to machinery, plants, and their maintenance. Also, many studies have been done using artificial neural networks for finding the output of the process. The study is also conducted using a deep learning method to forecast the output of renewable energy (Huang & Koroteev, 2021).

3.3.7 Unlawful discarding and waste removal

When people and companies discard waste in an area that is not designated for dumping, it is known as illegal dumping. It creates problems for society, the environment, and affects the health of the community. Waste generation and illegal dumping are increasing annually posing severe problems for the environment. All over the world, governments and policymakers must take strict action and make strategies to solve the problem (Niyobuhungiro & Schenck, 2021). Municipalities also play a major role in controlling illegal dumping. The installation of cameras and patrolling by personnel are also not yielding decent results in reducing the problem. AI can play a major role in finding the illegal dumping. The detection of illegal dumping can be increased by using a neural network model (Kim & Cho, 2022). The distance between the garbage bag and the dumper can be detected using AI technology to determine whether it is illegal dumping (Takahashi et al., 2022). Also single-shot detector algorithm can be used to find illegal dumping places by utilizing the photos taken by the drones (Youme et al., 2021). Also, a random forest algorithm, which is a technology of AI, can be used to find illegal dumping places (Ulloa-Torrealba et al., 2023).

3.3.8 Future scope

The importance of AI in waste management is increasing in the present scenario. The future scope in this field is vast and evolving day by day. The association between AI, IoT, and machine learning will increase competence in the future (Mousavi et al., 2023; Arora Madhu et al., 2025). There can be collaboration and knowledge sharing between different stakeholders like waste management companies, policymakers, technology providers, and researchers to increase the overall productivity. By implementing AI-driven technology for information and knowledge sharing, stakeholders can connect, share intellect, improve the innovative waste management system, and ensure continuous development in waste management (Amirsoleymani et al., 2022). Implementing policy and regulatory frameworks is highly required for AI to promote data allocation, incentivizing those adopting AI for waste management, innovation, and sustainability. Regulatory authorities should essentially remain active in upgrading the policies according to the requirements to solve the problems and issues in the existing system (Singh et al., 2022).

3.4 Conclusion

There should be collaboration and knowledge sharing between different stakeholders like waste management companies, policymakers, technology providers, and researchers to solve the data security issues associated with the AI-driven waste management system. Plans should be implemented to confirm the accessibility and correctness of information, which is the main requirement of an AI-driven system. The stakeholders should maintain high-quality data and confidentiality for better prediction and decision-making. A balance should be maintained between AI and the privacy and security of sensitive information. Governments, organizations, and waste management agencies must invest in the infrastructure required for the AI-driven waste management system. It requires smart sensors, connectivity, and IoT devices for real-time data collection and proper operation of the system. There should be continuous research in the sustainable AI-driven waste management process. The personnel involved with AI-driven waste management systems must receive proper training to enhance their knowledge and capacity building. Ethical and social considerations associated with AI in waste management, such as job dislocation, conservational justice, and municipal engagement, are required to be addressed. There should be discussions involving ethicists, officials, municipal representatives, policymakers, and government and industry experts for the responsible adoption of AI.

References

Al-Ruzouq, R., Abdallah, M., Shanableh, A., Alani, S., Obaid, L., & Gibril, M. B. A. (2022). Waste to energy spatial suitability analysis using hybrid multi-criteria machine learning approach. *Environmental Science and Pollution Research*, *29*, 2613-2628.

Ali, R. A., Nik Ibrahim, N. N. L., Wan Ab Karim Ghani, W. A., Lam, H. L., & Sani, N. S. (2021). Utilization of process network synthesis and machine learning as decision-making tools for municipal solid waste management. *International Journal of Environmental Science and Technology*, 1-12.

Altin, F. G., Budak, İ., & Özcan, F. (2023). Predicting the amount of medical waste using kernel-based SVM and deep learning methods for a private hospital in Turkey. *Sustainable Chemistry and Pharmacy*, *33*, 101060.

Amirsoleymani, Y., Abessi, O., & Ghajari, Y. E. (2022). A spatial decision support system for municipal solid waste landfill sites (case study: The Mazandaran Province, Iran). *Waste Management & Research*, *40*(7), 940-952.

Andeobu, L., Wibowo, S., & Grandhi, S. (2022). Artificial intelligence applications for sustainable solid waste management practices in Australia: A systematic review. *Science of The Total Environment*, *834*, 155389.

Baduge, S. K., Thilakarathna, S., Perera, J. S., Arashpour, M., Sharafi, P., Teodosio, B., ... & Mendis, P. (2022). Artificial intelligence and smart vision for building and construction 4.0: Machine and deep learning methods and applications. *Automation in Construction*, *141*, 104440.

Bernat, K. (2023). Post-consumer plastic waste management: From collection and sortation to mechanical recycling. *Energies*, *16*(8), 3504.

Bijos, J. C., Zanta, V. M., Morató, J., Queiroz, L. M., & Oliveira-Esquerre, K. P. (2022). Improving circularity in municipal solid waste management through machine learning in Latin America and the Caribbean. *Sustainable Chemistry and Pharmacy*, *28*, 100740.

Chauhan, R., Shighra, S., Madkhali, H., Nguyen, L., & Prasad, M. (2023). Efficient future waste management: A learning-based approach with deep neural networks for smart system (LADS). *Applied Sciences*, *13*(7), 4140.

Chen, L., Huang, L., Hua, J., Chen, Z., Wei, L., Osman, A. I., ... & Yap, P. S. (2023). Green construction for low-carbon cities: a review. *Environmental chemistry letters*, *21*(3), 1627-1657.

Chidepatil, A., Bindra, P., Kulkarni, D., Qazi, M., Kshirsagar, M., & Sankaran, K. (2020). From trash to cash: how blockchain and multi-sensor-driven artificial intelligence can transform a circular economy of plastic waste?. *Administrative Sciences*, *10*(2), 23.

Chiu, M. C., Wen, C. Y., Hsu, H. W., & Wang, W. C. (2022). Key wastes selection and prediction improvement for biogas production through hybrid machine learning methods. *Sustainable Energy Technologies and Assessments*, *52*, 102223.

Cubillos, M. (2020). Multi-site household waste generation forecasting using a deep learning approach. *Waste Management*, *115*, 8-14.

Edinov, S., & Fauzi, R. (2023). Community behavior in artificial intelligence-based waste management. *Formosa Journal of Sustainable Research*, *2*(2), 341-350.

Farghali, M., Osman, A. I., Mohamed, I. M., Chen, Z., Chen, L., Ihara, I., ... & Rooney, D. W. (2023). Strategies to save energy in the context of the energy crisis: a review. *Environmental Chemistry Letters*, *21*(4), 2003-2039.

Feng, Z., Yang, J., Chen, L., Chen, Z., & Li, L. (2022). An intelligent waste-sorting and recycling device based on improved EfficientNet. *International Journal of Environmental Research and Public Health*, *19*(23), 15987.

Guo, H. N., Wu, S. B., Tian, Y. J., Zhang, J., & Liu, H. T. (2021). Application of machine learning methods for the prediction of organic solid waste treatment and recycling processes: A review. *Bioresource technology*, *319*, 124114.

Huang, J., & Koroteev, D. D. (2021). Artificial intelligence for planning of energy and waste management. *Sustainable Energy Technologies and Assessments*, *47*, 101426.

Huang, L., Song, T., & Jiang, T. (2023). Linear regression combined KNN algorithm to identify latent defects for imbalance data of ICs. *Microelectronics Journal, 131*, 105641.

Hussain, A., Draz, U., Ali, T., Tariq, S., Irfan, M., Glowacz, A., ... & Rahman, S. (2020). Waste management and prediction of air pollutants using IoT and machine learning approach. *Energies, 13*(15), 3930.

Hu, X., Zhou, Y., Vanhullebusch, S., Mestdagh, R., Cui, Z., & Li, J. (2022). Smart building demolition and waste management frame with image-to-BIM. *Journal of Building Engineering, 49*, 104058.

Jassim, M. S., Coskuner, G., Sultana, N., & Hossain, S. Z. (2023). Forecasting domestic waste generation during successive COVID-19 lockdowns by Bidirectional LSTM super learner neural network. *Applied Soft Computing, 133*, 109908.

Jha, R., Dwivedi, S., & Modhera, B. (2022). Measurement and practices for hazardous waste management. In *Hazardous waste management* (pp. 89-115). Elsevier.

Kim, Y., & Cho, J. (2022). AIDM-Strat: augmented illegal dumping monitoring strategy through deep neural network-based spatial separation attention of garbage. *Sensors, 22*(22), 8819.

Krishna, G., & Sharma, A. (2023, February). A fuzzy logical based artificial intelligence method for designed to effectively predict and manage the solid waste. In *2023 IEEE International Conference on Integrated Circuits and Communication Systems (ICICACS)* (pp. 1-6). IEEE.

Lin, K., Zhao, Y., Kuo, J. H., Deng, H., Cui, F., Zhang, Z., ... & Wang, T. (2022). Toward smarter management and recovery of municipal solid waste: A critical review on deep learning approaches. *Journal of Cleaner Production, 346*, 130943.

Mao, W. L., Chen, W. C., Wang, C. T., & Lin, Y. H. (2021). Recycling waste classification using optimized convolutional neural network. *Resources, Conservation and Recycling, 164*, 105132.

Massoud, M. A., Abdallah, C., Merhbi, F., Khoury, R., & Ghanem, R. (2023). Development and application of a prioritization and rehabilitation decision support tool for uncontrolled waste disposal sites in developing countries. *Integrated Environmental Assessment and Management, 19*(2), 436-445.

Mousavi, S., Hosseinzadeh, A., & Golzary, A. (2023). Challenges, recent development, and opportunities of smart waste collection: A review. *Science of The Total Environment, 886*, 163925.

Munir, M. T., Li, B., & Naqvi, M. (2023). Revolutionizing municipal solid waste management (MSWM) with machine learning as a clean resource: Opportunities, challenges and solutions. *Fuel, 348*, 128548.

Niyobuhungiro, R. V., & Schenck, C. J. (2021). The dynamics of indiscriminate/illegal dumping of waste in Fisantekraal, Cape Town, South Africa. *Journal of Environmental Management, 293*, 112954.

Arora, M., Ahmad, V., Kumar, R., & Singh, R. (2025). The Impact of Industry 4.0 Technologies on Operational Excellence Methods: Application of Fuzzy Topsis. Engineering Management Journal, 1-12.

Ahmad, V., Arora, M., Sayal, A., & Kumar, R. Exploring Trends in E-Business and Sustainable Finance: An Integrated Analytical Approach. Frontiers in Sustainable Cities, 7, 1549874.

Devi, P., Desai, S. S., Kumar, R., Naruka, M. S., & Tripathi, V. (2024, December). A Strategic Innovations for Stock Market Optimization by Using Data Science & Explainable AI in High Frequency Trading. In 2024 International Conference on IoT Based Control Networks and Intelligent Systems (ICICNIS) (pp. 1510-1515). IEEE.

Kumar, R., Goel, R., Singh, T., Mohanty, S. M., Gupta, D., Alkhayyat, A., & Khanna, R. (2024). Sustainable Finance Factors in Indian Economy: Analysis on Policy of Climate Change and Energy Sector. Fluctuation & Noise Letters, 23(2).

Olawade, D. B., Wada, O. Z., Ore, O. T., David-Olawade, A. C., Esan, D. T., Egbewole, B. I., & Ling, J. (2024). Trends of solid waste generation during COVID-19 Pandemic: A review. *Waste Management Bulletin, 1*(4), 93-103.

Olawade, D. B., Fapohunda, O., Wada, O. Z., Usman, S. O., Ige, A. O., Ajisafe, O., & Oladapo, B. I. (2024). Smart waste management: A paradigm shift enabled by artificial intelligence. *Waste Management Bulletin*.

Osman, A. I., Hosny, M., Eltaweil, A. S., Omar, S., Elgarahy, A. M., Farghali, M., ... & Akinyede, K. A. (2023). Microplastic sources, formation, toxicity and remediation: a review. *Environmental Chemistry Letters, 21*(4), 2129-2169.

Pallathadka, H., Mustafa, M., Sanchez, D. T., Sajja, G. S., Gour, S., & Naved, M. (2023). Impact of machine learning on management, healthcare and agriculture. *Materials Today: Proceedings, 80*, 2803-2806.

Patel, A. K., Singhania, R. R., Albarico, F. P. J. B., Pandey, A., Chen, C. W., & Dong, C. D. (2022). Organic wastes bioremediation and its changing prospects. *Science of the Total Environment, 824*, 153889.

Pence, I., Kumaş, K., Siseci, M. C., & Akyüz, A. (2023). Modeling of energy and emissions from animal manure using machine learning methods: the case of the Western Mediterranean Region, Turkey. *Environmental Science and Pollution Research*, *30*(9), 22631-22652.

Peng X, Jiang Y, Chen Z, Osman AI, Farghali M, Rooney DW, Yap P-S (2023) Recycling municipal, agricultural and industrial waste into energy, fertilizers, food and construction materials, and economic feasibility: a review. Environ Chem Lett. https://doi.org/10.1007/s10311-022-01551-5

Pluskal, J., Šomplák, R., Nevrlý, V., Smejkalová, V., & Pavlas, M. (2021). Strategic decisions leading to sustainable waste management: Separation, sorting and recycling possibilities. *Journal of Cleaner Production*, *278*, 123359.

Ramya, P., Ramya, V., & Babu Rao, M. (2024). Optimized deep learning-based e-waste management in IoT application via energy-aware routing. *Cybernetics and Systems*, *55*(8), 2041-2070.

Reza, M., & Hassan, M. (2023). AI-Driven solutions for enhanced waste management and recycling in urban areas. *International Journal of Sustainable Infrastructure for Cities and Societies*, *8*(2), 1-13.

Salman, M. Y., & Hasar, H. (2023). Review on environmental aspects in smart city concept: Water, waste, air pollution and transportation smart applications using IoT techniques. *Sustainable Cities and Society*, *94*, 104567.

Singh, T., & Uppaluri, R. V. S. (2022). Machine learning tool-based prediction and forecasting of municipal solid waste generation rate: a case study in Guwahati, Assam, India. *International Journal of Environmental Science and Technology*, *20*(11), 12207-12230.

Sharma, P., & Vaid, U. (2021, November). Emerging role of artificial intelligence in waste management practices. In *IOP Conference Series: Earth and Environmental Science* (Vol. 889, No. 1, p. 012047). IOP Publishing.

Solano Meza, J.K., Orjuela Yepes, D., Rodrigo-Ilarri, J., Rodrigo-Clavero, M.-E., (2023). Comparative analysis of the implementation of support vector machines and long short-term memory artificial neural networks in municipal solid waste management and sustainability. Curr. Opin. Environ. Sci. Health 33, 100470. https://doi.org/10.1016/j.coesh.2023.100470.

Subramanian, A. K., Thayalan, D., Edwards, A. I., Almalki, A., & Venugopal, A. (2021). Biomedical waste management in dental practice and its significant environmental impact: A perspective. *Environmental Technology & Innovation*, *24*, 101807.

Tawfik, A., Mohsen, M., Ismail, S., Alhajeri, N. S., Osman, A. I., & Rooney, D. W. (2022). Methods to alleviate the inhibition of sludge anaerobic digestion by emerging contaminants: a review. *Environmental Chemistry Letters*, *20*(6), 3811-3836.

Takahashi Y, Fujii J, Amakata M (2022) Training a robust uav river patrol ai for diferent river and an analysis of the training dataset. In: Proceedings of the National Congress of the Society for Artifcial Knowledge. JSAI2022: 3S3IS2e01–3S3IS2e01.

Thaseen Ikram, S., Mohanraj, V., Ramachandran, S., & Balakrishnan, A. (2023). An intelligent waste management application using IoT and a genetic algorithm–fuzzy inference system. *Applied Sciences*, *13*(6), 3943.

Tsui, T. H., van Loosdrecht, M. C., Dai, Y., & Tong, Y. W. (2023). Machine learning and circular bioeconomy: Building new resource efficiency from diverse waste streams. *Bioresource Technology*, *369*, 128445.

Ulloa-Torrealba, Y. Z., Schmitt, A., Wurm, M., & Taubenböck, H. (2023). Litter on the streets-solid waste detection using VHR images. *European Journal of Remote Sensing*, *56*(1), 2176006.

Wang, X., Li, C., Lam, C. H., Subramanian, K., Qin, Z. H., Mou, J. H., ... & Lin, C. S. K. (2022). Emerging waste valorisation techniques to moderate the hazardous impacts, and their path towards sustainability. *Journal of hazardous materials*, *423*, 127023.

Wilts, H., Garcia, B. R., Garlito, R. G., Gómez, L. S., & Prieto, E. G. (2021). Artificial intelligence in the sorting of municipal waste as an enabler of the circular economy. *Resources*, *10*(4), 28.

Wu, T. W., Zhang, H., Peng, W., Lü, F., & He, P. J. (2023). Applications of convolutional neural networks for intelligent waste identification and recycling: A review. *Resources, Conservation and Recycling*, *190*, 106813.

Yadav, V., & Karmakar, S. (2020). Sustainable collection and transportation of municipal solid waste in urban centers. *Sustainable Cities and Society*, *53*, 101937.

Yan, B., Liang, R., Li, B., Tao, J., Chen, G., Cheng, Z., ... & Li, X. (2021). Fast identification and characterization of residual wastes via laser-induced breakdown spectroscopy and machine learning. *Resources, Conservation and Recycling*, *174*, 105851.

Youme, O., Bayet, T., Dembele, J. M., & Cambier, C. (2021). Deep learning and remote sensing: detection of dumping waste using UAV. *Procedia computer science*, *185*, 361-369.

Zhang, Q., Yang, Q., Zhang, X., Bao, Q., Su, J., & Liu, X. (2021). Waste image classification based on transfer learning and convolutional neural network. *Waste Management*, *135*, 150-157.

4

AI-driven Waste Minimization Model in Manufacturing

Priyanka Agrawal

Graphic Era deemed to be University, India
E-mail: priyankaagrawal.pdp@geu.ac.in

Abstract

The chapter investigates the potential use of AI-based waste minimization models in manufacturing by reviewing literature on their application and identifying gaps in the research on this topic. The goal of the presentation is to synthesize studies for knowledge development on the applications of AI in waste reduction and to formulate areas for relevant future work. The methodology of a systematic literature review was used to analyze the peer-reviewed studies retrieved from the reputable databases. The focus of the data analysis was on key AI models, including machine learning (ML), deep learning (DL), and predictive analytics, and the extent to which they have contributed to waste reduction in various manufacturing sectors. As a result, the findings show that in process optimization, defect detection, and predictive maintenance, AI models contribute significantly to reduce the material waste, energy consumption, and overproduction. However, there are the challenges such as high implementation costs, data quality issues, and scalability. Though AI has the potential to transform sustainable manufacturing, the study concludes that further research is required to improve accessibility, scalability, and integration into smaller enterprises.

Keywords: AI-driven waste minimization, machine learning, circular economy, predictive maintenance, deep learning, resource efficiency.

DOI: 10.1201/9788743806905-4

4.1 Introduction

Waste minimization in manufacturing is an area of focus due to the growing pressure on industries to meet sustainability goals. Waste in materials, energy and defective products accounts for the majority of resource use and waste generation in manufacturing. It is important to address this challenge, as it will help reduce the environmental footprint and improve operational efficiency. As the concern for the environmental sustainability has been growing, the significant shift in the approach towards the manufacture of goods shifted towards the sustainable manufacturing practice where there takes a heavy weight in the reduction of waste and the maximization of the resources utilized without reducing the performance of the product. However, it is well proven that AI technologies are the key drivers for this transformation. In recent years, AI-driven models have been gaining much popularity in several manufacturing industry sectors such as automotive, electronics, and food processing which are still traditional energy and material consumers. Khan et al. (2022) mention that AI models such as machine learning (ML), deep learning (DL), and predictive analytics are being used to optimize production schedules, detect issues early, and predict equipment failures, thereby decreasing material and energy waste. Particularly, integration of AI with IoT (Internet of Things) has come in handy in reducing waste. According to Garcia-Muina et al. (2018) AI and IoT integration enables continuous monitoring of the manufacturing process to facilitate the real-time changes in resource allocation and consumption of energy. This study aims to synthesize existing knowledge on AI applications for waste reduction in manufacturing and to identify knowledge gaps. Notably, the study will also examine the need for additional research on the integration of AI with IoT for waste tracking, as well as the impact of predictive maintenance in reducing downtime and energy consumption (Khan et al., 2022). These gaps need to be identified so that AI technologies can be advanced, which can be scalable and profitable in other manufacturing sectors. The aim of the study is to offer valuable insights for researchers, manufacturers, and policymakers to deploy AI-driven waste minimization models for improving the sustainability goals and manufacturing practices.

4.2 Review of Literature

4.2.1 Overview of waste minimization in manufacturing

Lean manufacturing and Six Sigma are two of the most well-known traditional approaches to waste minimization in manufacturing. These methodologies have been widely adopted by industries seeking to improve operational efficiency and reduce waste through systematic process optimization and quality control techniques. Kristoffersen et al. (2020) discuss how Lean principles are foundational in manufacturing, with an emphasis on waste reduction and continuous improvement. This focus aligns with AI-driven systems, which optimize production workflows and eliminate inefficiencies in real-time, enhancing traditional Lean methods (Sarc et al., 2019). Kerin et al. (2019) discuss the application of Six Sigma techniques, particularly in high-precision manufacturing environments, where maintaining consistency and minimizing variability are essential to reduce material waste. These traditional approaches are evolving with Industry 4.0 technologies as mentioned by Garcia-Muina et al. (2018). For example, AI-based predictive maintenance can prevent equipment failure, which is one of the common sources of waste in traditional systems, thereby improving efficiency and reducing downtime (Kerin et al., 2019). These methodologies are still of great importance in speeding up efficiency and waste reduction, as we are constantly experiencing the need for greater sustainability and operational excellence in a market place industry.

4.2.2 AI technologies in manufacturing

Now ML, DL, and predictive analytics are prevalent engines of modern manufacturing processes. According to Bressanelli et al. (2019), they demonstrated how the machine learning algorithms are used in predictive models for their manufacturing schedules optimization that resulted in improving production efficiency and lowering waste. Maiurova et al. (2019) define the use case of deep learning in automated defect detection. The use of AI models to detect defects in products during production enables prompt corrective actions, before the products go farther down the line.

4.2.3 AI applications in waste minimization

The use of AI applications has become a basic aspect in waste minimization strategies in the manufacturing area. AI uses machine learning (ML), deep learning (DL), and predictive analytics to streamline production processes, control defects, efficiently utilize resources, and, last but not least, predict when maintenance needs will arise. In their 2019 discussion, Khan et al. (2022) share that AI-based optimization algorithms should improve production efficiency by adjusting workflows using real-time data. Magrini et al. (2020) demonstrate how deep learning models are utilized in computer vision systems that detect product defect in real time in a production.

4.2.4 AI-driven waste minimization models

With the use of AI-driven models, processes are optimized, defects are detected, and resource management is improved, leading to waste minimization in manufacturing. Machine learning (ML), deep learning (DL), and optimization are classified algorithms utilized by these AI models to minimize waste on various stages of the manufacturing process. Some prominent applied waste minimization models based on AI are as follows.

4.2.4.1 ML-based defect detection models

Defect detection in manufacturing is widely automated through machine learning models, especially when defect knowledge is provided in advance. According to Alves et al. (2020), machine learning algorithms play an important role in detecting defects early in the production process. Bressanelli et al. (2020) mention that even though these models can be extremely effective, training on large volumes of high-quality data is required, especially in environments where data collection is sparse and spurious.

4.2.4.2 Optimization algorithms for process efficiency

Manufacturing requires the use of optimization algorithms as key AI models to reduce waste and enhance process efficiency. Ingemarsdotter et al. (2019) suggest the usage of genetic algorithms and simulated annealing for optimizing production schedules and resource allocation. While these provide powerful insights, Alves et al. (2020) show that their implementation in complex manufacturing environments demands significant computational resources and expertise, potentially creating a barrier for smaller manufacturers.

4.2.4.3 AI for resource management

Another important area where AI is applied to minimize waste is resource management. The work of Del Giudice et al. (2020) investigates how AI-based models are used in supply chain management to predict the required amount of raw materials, thereby preventing overstocking and reducing waste from expired or unused materials. As per Del Giudice et al. (2020), these models depend strongly on the data quality and accuracy from production systems and sensors. If your data is insufficient or ill conceived, the predictions will be faulty and might, in the end, contribute to unnecessary waste or inefficiency.

4.2.4.4 Predictive maintenance models

Other applications of AI in waste minimization are predictive maintenance models. In these models, real-time data from manufacturing equipment is analyzed to determine when the equipment is likely to fail. A major source of downtime and waste in production processes, as discussed by Magrini et al. (2020), is machine breakdowns, and predictive analytics play a role in preventing them.

4.2.4.5 Effectiveness and limitations of AI models

The implementation and long-term success of AI models depend on several factors such as data quality, system integration, and computing resources. In fact, just like the authors of Bhubalan et al. (2020) notice, the successful use of AI models in real manufacturing settings heavily relies on having access to high-quality real-time data, which might not always be available in every manufacturing setting. As Magrini et al. (2020) explain, AI-enabled systems are being employed to produce this loop in production, increasing the recycling and reuse of materials.

4.3 Research Objectives

This gap analysis aims to scrutinize the body of literature about AI-driven waste minimization in manufacturing. Important tasks involve studying AI models like machine learning (ML) and deep learning (DL), and their use in various manufacturing subsectors such as automotive, electronics, and food processing industries. Some limitations like data quality, costs, scalability, and integration will also be analyzed. The primary research question that this study attempts to answer is, "What is the state of the art of waste minimization models in manufacturing from an AI-driven perspective, and what gaps are there that need to be addressed?" This question aims to motivate efforts toward providing solutions to improve the role AI plays in sustainability and waste minimization in the future.

4.4 Research methodology

This research undertakes a systematic literature review (SLR) to assess the impact of technological innovations on the development of a circular economy (CE). The analysis incorporates 46 scientific publications collected from Scopus, Web of Science, and Google Scholar databases, which guarantee a complete and high-standard dataset. The methodology is based on a predetermined approach defined by the stages of search strategy, inclusion criteria, and data extraction aimed to reveal relevant technological aspects, issues, and policies regarding CE.

4.4.1 Search query used

(TITLE-ABS-KEY("circular economy" OR "closed-loop economy" OR "sustainable economy"OR"green economy" OR "regenerative economy")) AND (TITLE-ABS-KEY("technological innovation"OR "digital transformation" OR "emerging technology*" OR "automation" OR "smart technology*" OR"digitalization")) AND (TITLE-ABS-KEY("artificial intelligence" OR "machine learning" OR"blockchain" OR "IoT" OR "big data" OR "cloud computing" OR "cyber-physical systems" OR"robotics")) AND (TITLE-ABS-KEY("resource efficiency" OR "waste management" OR "circularsupply chain*" OR "sustainable supply chain*" OR "remanufacturing" OR "recycling technology*" OR"eco-efficiency"))

4.4.2 Inclusion and exclusion criteria

This study required the selection of AI-driven waste minimization in manufacturing peer-reviewed articles only. Ensuring methodological quality, only articles published from 2013 to 2023 were selected. The papers examined the role of artificial intelligence (AI) technologies like machine learning (ML), deep learning (DL), Internet of Things (IoT), big data, and predictive analytics in resource and waste minimization and efficiency in the manufacturing domain. Priority was given to articles that explained the application of AI in process optimization, defect detection, resource management, and predictive maintenance in manufacturing.

4.5 Systematic Review and Gap Analysis

The studies present the need for AI applications in detection of the defective products, optimization of processes, management of resources, and predictive maintenance that will all help in cutting down on wastes. For instance, AI applications in the automotive and electronics industries have been most useful in material waste minimization, energy management, and materials scheduling.

Table 4.1 Systematic analysis

Study	AI models/ technologies	Waste minimization strategy	Industry/ sectors	Key findings	Limitations/ gaps
Kristoffersen et al. (2020)	IoT, AI algorithms	Process optimization, energy efficiency	Automotive, electronics	AI and IoT integration improves energy efficiency, reduces material waste through real-time data monitoring	High cost of AI implementation in existing manufacturing systems
Sarc et al. (2019)	Machine learning (ML), robotics	Defect detection, process automation	Manufacturing, automotive	Robotics and AI enhance quality control, detect defects early, minimize scrap and waste in production processes	Difficulty in scaling AI for small enterprises
Kerin et al. (2019)	Deep learning (DL), predictive maintenance	Reducing downtime, resource optimization	Electronics, automotive	Predictive maintenance reduces unplanned downtime, optimizing equipment usage and reducing waste from inefficiency	Requires high-quality sensor data to be effective
Del Giudice et al. (2020)	Machine learning (ML), predictive analytics	Resource management, energy consumption	General manufacturing	AI improves energy usage efficiency and resource allocation, reducing material waste	Models require frequent recalibration to remain accurate
Garcia-Muina et al. (2018)	IoT, big data, AI	Real-time monitoring, waste reduction	Electronics, manufacturing	IoT and AI systems track resources and processes, reducing material waste and improving sustainability	High initial costs and integration challenges
Rajput et al. (2020)	Machine learning, AI for resource efficiency	Overproduction reduction, waste minimization	Electronics	AI models help predict demand, preventing overproduction and material waste	Difficulty in adapting models for diverse manufacturing setups

(Continued.)

Table 4.1 *Continued.*

Study	AI models/ technologies	Waste minimization strategy	Industry/ sectors	Key findings	Limitations/ gaps
Bressanelli et al. (2020)	Optimization algorithms, AI	Process optimization, waste minimization	Manufacturing	Optimization algorithms help streamline processes, reducing energy consumption and material waste	Requires large datasets for training and effective model use
Maiurova et al. (2019)	Machine learning, deep learning	Defect detection, process optimization	Automotive	ML and DL models detect defects in real-time, reducing waste due to faulty products	Data imbalance and quality issues affect model accuracy
Ingemarsdotter et al. (2019)	Deep learning (DL), IoT	Defect detection, quality control	Automotive, electronics	AI and IoT integration for real-time defect detection reduces scrap and improves product quality	High costs for sensor systems and data infrastructure
Reuter et al. (2019)	Predictive analytics, machine learning	Process efficiency, waste minimization	Automotive, manufactur-ing	Predictive analytics forecast process bottlenecks, optimizing resource usage and minimizing waste	Complexity of integrating predictive models with existing systems
Kerin et al. (2020)	IoT, deep learning, robotics	Automation, waste reduction	Automotive, electronics	AI-driven robotics enhance production efficiency, reducing waste and improving recycling	Implementation cost and integration issues in traditional factories
Sniatala et al. (2020)	Machine learning, AI	Waste reduction, process optimization	General man-ufacturing	AI models optimize production schedules, reducing overproduction and energy waste	Inconsistent data quality can affect the model's performance
Kurniawan et al. (2020)	Machine learning, AI	Resource management, waste minimization	Electronics	AI for resource management improves material allocation, reducing resource wastage	Lack of universal AI models for different industries

Table 4.1 *Continued.*

Study	AI models/ technologies	Waste minimization strategy	Industry/ sectors	Key findings	Limitations/ gaps
Alves et al. (2021)	AI, IoT, blockchain	A circular economy, resource recovery	Electronics, automotive	AI and blockchain improve material traceability in circular supply chains, reducing waste	High initial investment required for integrating blockchain with AI
Aziz et al. (2021)	AI, robotics, deep learning	Automation, waste reduction	Manufacturing	AI-driven robots improve efficiency and reduce waste by automating repetitive tasks	Requires high capital investment and skilled labor for setup
Elghaish et al. (2020)	Machine learning, predictive analytics	Energy efficiency, waste reduction	Automotive	Predictive models optimize energy consumption, leading to waste minimization in production processes	Predictive models require frequent updates for accuracy
Ghoreishi et al. (2020)	AI, robotics, IoT	Waste reduction, resource efficiency	Automotive, electronics	IoT and robotics improve production efficiency and reduce waste by providing real-time process feedback	Difficulty in achieving seamless integration across multiple platforms
Srivastav et al. (2023)	Machine learning, robotics	Process automation, waste minimization	Manufacturing	AI-driven automation and robotics optimize manufacturing workflows, reducing material waste	Lack of standardized AI solutions for different production environments
Said et al. (2023)	Deep learning, predictive analytics	Defect detection, quality control	Manufacturing	Deep learning enhances defect detection accuracy, significantly reducing scrap and rework	Over-reliance on AI for quality control can result in errors
Chen et al. (2022)	Machine learning, AI	Process optimization, defect detection	Electronics, automotive	AI-driven models predict and optimize manufacturing schedules, reducing waste from defective products	High computational resources required for real-time data processing

Table 4.1 *Continued.*

Study	**AI models/ technologies**	**Waste minimization strategy**	**Industry/ sectors**	**Key findings**	**Limitations/ gaps**
Khadke et al. (2021)	Blockchain, AI	Plastic recycling, circular supply chain	Electronics, manufacturing	AI and blockchain improve plastic waste management through a closed-loop recycling process	Requires large-scale infrastructure and data standardization
Magrini et al. (2021)	IoT, AI, blockchain	Waste management, a circular economy	Electronics, manufacturing	IoT and blockchain enhance traceability and resource efficiency in circular supply chains	High implementation cost and lack of interoperability between technologies
Jiang et al. (2023)	Blockchain, AI	Waste management, resource efficiency	Manufacturing, electronics	Blockchain-integrated waste management improves traceability and encourages recycling in waste management systems	Scalability in larger operations is challenging, requiring extensive data networks
Bhubalan et al. (2021)	AI, blockchain	Resource recovery, circular supply chain	Manufacturing, automotive	Blockchain and AI enable efficient resource recovery and tracking in the circular supply chain	Limited research on the real-time integration of blockchain with AI in waste management
Kroell et al. (2020)	AI, digital technologies	Waste reduction, recycling	Electronics, automotive	Digital technologies enable efficient recycling by enhancing monitoring and optimization	High initial costs and need for skilled workforce for implementation
Cwiklicki et al. (2020)	AI, Industry 4.0	Waste minimization, process automation	General manufacturing	Industry 4.0 technologies, including AI, automate processes and reduce material waste through real-time process adjustments	Dependency on large datasets for training and complex integration with legacy systems

Table 4.1 *Continued.*

Study	AI models/ technologies	Waste minimization strategy	Industry/ sectors	Key findings	Limitations/ gaps
Alonso et al. (2021)	AI, robotics	Waste reduction, process efficiency	Electronics	AI-driven robotics in production lines reduces material waste and increases production efficiency	Implementation complexity in nonautomated industries and high upfront costs
Tavana et al. (2021)	AI, ML	Resource management, waste minimization	Automotive, aerospace	Machine learning models predict resource needs, optimizing material flow and reducing overproduction	Requires consistent, high-quality data for optimal performance
Pakseresht et al. (2021)	AI, predictive analytics	Waste minimization, process optimization	Manufacturing	Predictive analytics models forecast demand and optimize production schedules, minimizing waste	Models require continuous recalibration to maintain accuracy
Rodrigo et al. (2023)	AI, big data	Waste reduction, process optimization	Manufacturing, food processing	Big data and AI improve waste reduction by optimizing production and minimizing spoilage	Limited research on real-time adaptability and integration across diverse manufacturing environments
Quayson et al. (2023)	AI, IoT	Energy optimization, waste minimization	Electronics, automotive	AI and IoT integration helps optimize energy consumption and reduce waste in manufacturing operations	High infrastructure cost for IoT sensors and AI integration
Islam et al. (2022)	AI, IoT	Waste minimization, resource efficiency	Manufacturing	IoT-enabled AI systems provide real-time feedback to minimize waste and enhance resource efficiency	Lack of standardization in IoT sensor data integration across platforms

Table 4.1 *Continued.*

Study	AI models/ technologies	Waste minimization strategy	Industry/ sectors	Key findings	Limitations/ gaps
Steenmans et al. (2022)	AI, robotics	Defect detection, waste reduction	Automotive	Robotics and AI enhance defect detection accuracy, preventing defective products from moving down the production line, reducing waste	High computational cost and need for robust training datasets
Khan et al. (2022)	AI, machine learning	Process optimization, defect detection	Electronics, automotive	Machine learning improves product quality by detecting defects early in the production process	Lack of scalability in small-scale manufacturing environments
Khan et al. (2022)	AI, deep learning	Defect detection, waste minimization	Automotive, electronics	Deep learning is applied for real-time defect detection, reducing waste and increasing product quality	Requires vast amounts of data for training and ongoing adjustments
Sharma et al. (2025)	AI, big data	Waste reduction, resource management	Electronics	Big data and AI optimize resource management by reducing waste and enhancing energy efficiency in manufacturing	Data inconsistency across manufacturing systems and challenges in integration
Shukla et al. (2022)	AI, IoT	Resource efficiency, waste minimization	Automotive	IoT-based AI systems enable real-time resource tracking and management, optimizing material use and reducing waste	High implementation costs and dependence on external IoT infrastructure
Chaouni et al. (2023)	Blockchain, AI	Waste reduction, circular supply chain	Manufacturing, electronics	Blockchain and AI improve material traceability and help close the loop in circular supply chains, enhancing waste reduction	Requires integration of various technological systems and standardized practices

Table 4.1 *Continued.*

Study	AI models/ technologies	Waste minimization strategy	Industry/ sectors	Key findings	Limitations/ gaps
Patil et al. (2023)	AI, robotics	Waste minimization, process automation	Manufacturing	AI-driven robotics in production lines help reduce scrap, improve efficiency, and minimize material waste	Initial cost of robotic systems can be prohibitive for SMEs
Nafiz et al. (2023)	AI, machine learning	Waste minimization, process optimization	Manufacturing	Machine learning models predict demand and optimize material flows to reduce production waste	Models require continuous recalibration to accommodate fluctuating demand
Govindan et al. (2024)	AI, big data	Resource management, energy optimization	Manufacturing	AI and big data improve resource allocation and optimize energy use, significantly reducing waste	Need for real-time data analytics and integration with legacy systems
Martikkala et al. (2023)	IoT, AI	Waste minimization, resource efficiency	Manufacturing	IoT and AI together help track resource use and waste production in real-time, improving sustainability	High cost of sensors and maintenance challenges for IoT systems
Singh et al. (2024)	AI, robotics	Waste reduction, automation	Manufacturing	AI and robotics reduce material waste by automating repetitive tasks and enhancing product quality	High complexity of AI systems limits their accessibility for smaller manufacturers
Kanyilmaz et al. (2023)	AI, machine learning	Waste reduction, defect detection	Manufacturing	ML models enhance defect detection, reducing scrap and waste in production lines	Data imbalance and insufficient training data in some manufacturing sectors

Table 4.1 *Continued.*

Study	**AI models/ technologies**	**Waste minimization strategy**	**Industry/ sectors**	**Key findings**	**Limitations/ gaps**
Kaiser et al. (2022)	AI, IoT, blockchain	A circular economy, resource recovery	Automotive, electronics	Blockchain and IoT integration with AI enhance traceability and improve the efficiency of recycling and waste recovery	Integration challenges in legacy systems and high technological investment costs
Singh et al. (2024)	AI, big data	Waste minimization, energy efficiency	Electronics, manufacturing	Big data and AI technologies optimize energy use in manufacturing, reducing material waste and energy consumption	Scalability issues in applying AI solutions to smaller operations

SOURCE: AUTHOR

4.6 Outcome of Systematic Analysis

By the systematic analysis of the literature, one observes the application of AI-driven models such as predictive analytics, machine learning (ML), and deep learning (DL) for minimizing waste in the manufacturing. These AI models are being refined to optimize processes, detect defects, and manage resources across industries such as automotive, electronics, and food processing. Further research is needed to make it more accessible and applicable to industries.

Table 4.2 Outcome of systematic analysis

Category	**Key findings**	**Challenges/limitations**	**Future research directions**
AI models	AI models such as ML, DL, and predictive analytics are widely used for defect detection, process optimization, and predictive maintenance	High implementation costs, especially in SMEs. Requirement for high-quality data for model training	Research into scalable AI solutions for small and medium-sized enterprises (SMEs)
Waste minimization strategies	AI-driven strategies include reducing overproduction, energy consumption, and material waste through real-time data and AI-powered optimization	Integration of AI into existing manufacturing systems is complex and requires substantial infrastructure	Investigation into how AI can reduce waste across varied industries and in small-scale operations

Table 4.2 *Continued.*

Category	Key findings	Challenges/limitations	Future research directions
Industry applications	The automotive, electronics, and food processing industries benefit the most from AI in waste reduction, particularly in resource management	Limited application of AI in industries outside automotive and electronics	Expand research into cross-industry applications and the global scalability of AI-driven waste reduction models
Challenges and limitations	Data quality and integration issues with legacy systems are major hurdles. AI models require vast datasets for optimal functioning	Data inconsistency across industries and integration challenges with legacy systems limit effectiveness	Development of standardized data protocols and integration frameworks for AI in manufacturing
Scalability and adaptability	Large-scale AI models often lack scalability for SMEs, making them less applicable to smaller manufacturing environments	High infrastructure and training costs. AI models designed for large operations often fail in smaller contexts	Research into low-cost, scalable AI solutions for small to medium-sized manufacturing environments
Technological integration	IoT and AI integration for real-time feedback on energy efficiency and resource usage is becoming increasingly popular	High initial costs for IoT sensors and AI setup. Challenges in the real-time adaptability of AI systems	Explore integration with other technologies like blockchain and big data to enhance effectiveness in waste reduction

SOURCE: AUTHOR

4.7 Case Study: AI-driven Waste Minimization in Agile Manufacturing using Recycled Materials

Background: In the context of environmental sustainability in manufacturing, there is a requirement to develop processes that minimize the waste and maximize the utilization of resources. Artificial intelligence (AI) integration into manufacturing can go a long way toward such targets.

Problems they solve: Manufacturers struggle with the challenges around incorporation of recycled materials into production keeping the product quality constant. Traditional methods do not provide the required precision when it comes to predicting and controlling the properties of recycled materials during manufacturing.

Varde and Liang (2023) suggested the use of machine learning techniques to help in decision-making for agile manufacturing with recycled materials.

Artificial neural networks (ANNs) were used by them to study the parameters influencing heat treatment process and their effect on material properties. Furthermore, for grain size detection, convolutional neural networks (CNNs) were used while phase fraction detection was examined with the use of random forest classifiers. The goal of these AI models was to predict and optimize how recycled materials might affect the outcomes of manufacturing processes.

Result: These machine learning models were successfully implemented. The accuracy of prediction of microstructural development during a quench tempering process, i.e., a heat treatment process, was about 90% using the ANN model. The impressive level of accuracy suggests that AI can successfully support manufacturing decision-making involving recycled materials.

Key learnings:

- Recycling manufacturing process: The prediction and optimization of manufacturing processes involving recycled materials can be greatly improved by integrating machine learning models (ANN, CNNs, and random forests).
- AI-driven analyses: This helps manufacturers optimize the processing parameters to improve material properties and reduce the waste.
- Prestige: Recycling materials decreases its raw material requirements and enhances its prestige as it can be recycled without sacrifice.

Reference:

Varde, A. S., & Liang, J. (2023). Machine Learning Approaches in Agile Manufacturing with Recycled Materials for Sustainability. arXiv preprint arXiv:2303.08291.

4.8 Findings

A systematic review and literature gap analysis of existing AI-based models for waste minimization in manufacturing has been conducted, revealing their efficacy. Khan et al. (2022). indicate that AI models have been able to optimize the production schedules, detect defects, and prevent unplanned downtime by means of predictive maintenance. In particular, the automotive and electronics sectors are cases in which AI applications decrease material waste, eliminate scrap, and optimize the use of the resources (Maiurova et al., 2019; Bressanelli et al., 2020). The other key point is increased importance of predictive analytics in resource management. As Garcɀa-Muīna et al. (2018) mention, AI-driven predictive models lead to more accurate demand forecasting, which in turn helps to do away with overproduction and reduce inventory waste. Despite this, challenges remain, including the high implementation cost of AI, especially for SMEs (Sarc et al., 2019). Moreover, there exist significant obstacles to data quality, which is itself a huge barrier to AI model performance (Kristoffersen et al., 2020).

4.8.1 Impact of AI on waste minimization

In manufacturing, waste minimization for material, energy consumption, and overproduction can be largely attributed to the impact of AI technologies. According to Garcia-Muina et al. (2018), AI-based defect detection systems offer excellent potential to detect defect very early in production cycle thereby resulting in waste as a result of rework and scrap, especially in automotive and other industries (Said et al., 2023). Furthermore, predictive analytics models assist manufacturers to minimize energy waste by optimizing energy use according to real-time data by adjusting the production schedules; the machinery consumes energy only when it is needed (Reuter et al., 2019). Secondly, the AI optimization models anticipate the demand and supply, thereby striking a balance between demand and supply and reducing waste caused by overproduction.

4.8.2 Emerging trends

Recently, several patterns of AI-driven waste minimization have been observed in manufacturing processes. The integration of AI with the circular economy (CE) is a large trend. Alves et al. (2021) note how AI is used to increase material recycling in closed-loop supply chains, waste reduction, and increase in resource efficiency by optimizing recycling processes. Furthermore, the combination of blockchain and AI is being employed to more efficiently monitor material supply chain operations, ensuring that all resources are reused and recycled, thereby reducing waste (Bhubalan et al., 2021). Ingemarsdotter et al. (2019) and Kanyilmaz et al. (2020) emphasize the importance of IoT-enabled AI systems that maintain continuous monitoring of the production process, and could track the waste in real time. At last, the integration of AI with big data and blockchain has shown great potential for improving resource efficiency.

4.9 Conclusion

Analysis of gap between literature review and the actuality of this area has placed both theoretical as well as practical value in the role of AI-driven waste minimization models in manufacturing. Therefore, the findings apply AI technologies like machine learning (ML), deep learning (DL), and predictive analytics into optimizing processes, defect detection, and predictive maintenance, which in turn would result in waste reduction. Waste in manufacturing can be reduced through the strategies of reducing over production, material waste, energy consumption, and resource management. Moreover, scalability is an issue that has been (and still is) a matter of concern, among other things in smaller and medium-sized enterprises (SMEs). Furthermore, many of of these AI models are designed for large-scale operations, so they still need further refinement to be generally applicable in different manufacturing environments. Besides reducing waste, innovations in AI also contribute to a more sustainable manufacturing future. Achieving long-term sustainability goals is essential for the future of the manufacturing industry, and it is this continued research and innovation that are critical.

References

Alonso S.L.N.; Forradellas R.F.R.; Morell O.P.; Jorge-Vazquez J. (2021), Digitalization, a circular economy and environmental sustainability: The application of artificial intelligence in the efficient self-management of waste : Sustainability (Switzerland), 13(4), 1-20 ; DOI Link:10.3390/su 13042092

Alves L.; Ferreira Cruz E.; Lopes S.I.; Faria P.M.; Rosado da Cruz A.M. (2022), Towards a circular economy in the textiles and clothing value chain through blockchain technology and IoT: A review : Waste Management and Research, 40(1), 3-23 ; DOI Link:10.1177/0734242X211052858

Aziz N.A.; Adnan N.A.A.; Wahab D.A.; Azman A.H. (2021), Component design optimisation based on artificial intelligence in support of additive manufacturing repair and restoration: Current status and future outlook for remanufacturing : Journal of Cleaner Production, 296, - ; DOI Link:10.1 016/j.jclepro.2021.126401

Bhubalan K.; Tamothran A.M.; Kee S.H.; Foong S.Y.; Lam S.S.; Ganeson K.; Vigneswari S.; Amirul A.-A.; Ramakrishna S. (2022), Leveraging blockchain concepts as watermarkers of plastics for sustainable waste management in progressing a circular economy : Environmental Research, 213, - ; DOI Link:10.1016/j.envres.2022.113631

Bressanelli G.; Pigosso D.C.A.; Saccani N.; Perona M. (2021), Enablers, levers and benefits of a Circular Economy in the Electrical and Electronic Equipment supply chain: a literature review : Journal of Cleaner Production, 298, - ; DOI Link:10.1016/j.jclepro.2021.126819

Bressanelli G.; Pigosso D.C.A.; Saccani N.; Perona M. (2021), Enablers, levers and benefits of a Circular Economy in the Electrical and Electronic Equipment supply chain: a literature review : Journal of Cleaner Production, 298, - ; DOI Link:10.1016/j.jclepro.2021.126819

Chaouni Benabdellah A.; Zekhnini K.; Cherrafi A.; Garza-Reyes J.A.; Kumar A.; El Baz J. (2023), Blockchain technology for viable circular digital supplychains: an integrated approach forÂ evaluating the implementation barriers : Benchmarking, 30(10), 4397-4424 ; DOI Link:10.1108/BIJ-04-2022-0240

Chen X. (2022), Machine learning approach for a circular economy with waste recycling in smart cities : Energy Reports, 8, 3127-3140 ; DOI Link:10.1016/j.egyr.2022.01.193

Cwiklicki M.; Wojnarowska M. (2020), A circular economy and industry 4.0: One-way or two-way relationships? : Engineering Economics, 31(4), 387-397 ; DOI Link:10.5755/j01.ee.31.4.24565

Del Giudice M.; Chierici R.; Mazzucchelli A.; Fiano F. (2020), Supply chain management in the era of a circular economy: the moderating effect of big data : International Journal of Logistics Management, 32(2), 337-356 ; DOI Link:10.1108/IJLM-03-2020-0119

Elghaish F.; Hosseini M.R.; Kocaturk T.; Arashpour M.; BararzadehLedari M. (2023), Digitalised circular construction supply chain: An integrated BIM-Blockchain solution : Automation in Construction, 148, - ; DOI Link: 10.1016/j.autcon.2023.104746

Ghoreishi M.; Happonen A. (2020), New promises AI brings into a circular economy accelerated product design: A review on supporting literature : E3S Web of Conferences, 158, - ; DOI Link:10.1051/e3sconf/202015806002

Govindan K. (2024), Unlocking the potential of quality as a core marketing strategy in remanufactured circular products: A machine learning enabled multi-theoretical perspective : International Journal of Production Economics, 269, - ; DOI Link:10.1016/j.ijpe.2023.109123

Ingemarsdotter E.; Jamsin E.; Kortuem G.; Balkenende R. (2019), Circular strategies enabled by the internet of things-a framework and analysis of current practice : Sustainability (Switzerland), 11(20), - ; DOI Link:10.3390/su11205689

Islam M.T.; Iyer-Raniga U.; Trewick S. (2022), Recycling Perspectives of Circular Business Models: A Review : Recycling, 7(5), - ; DOI Link:10.3390/recycling7050079

Jiang P.; Zhang L.; You S.; Fan Y.V.; Tan R.R.; Kleme J.J.; You F. (2023), Blockchain technology applications in waste management: Overview, challenges and opportunities : Journal of Cleaner Production, 421, - ; DOI Link:10.1016/j.jclepro.2023.138466

Kaiser J.-P.; Lang S.; Wurster M.; Lanza G. (2022), A Concept for Autonomous Quality Control for Core Inspection in Remanufacturing : Procedia CIRP, 105, 374-379 ; DOI Link:10.1016/j.procir.2022.02.062

Kanyilmaz A.; Birhane M.; Fishwick R.; del Castillo C. (2023), Reuse of Steel in the Construction Industry: Challenges and Opportunities : International Journal of Steel Structures, 23(5), 1399-1416 ; DOI Link: 10.1007/s13296-023-00778-4

Kerin M.; Pham D.T. (2019), A review of emerging industry 4.0 technologies in remanufacturing : Journal of Cleaner Production, 237, - ; DOI Link: 10.1016/j.jclepro.2019.117805

Kerin M.; Pham D.T. (2020), Smart remanufacturing: a review and research framework : Journal of Manufacturing Technology Management, 31(6), 1205-1235 ; DOI Link:10.1108/JMTM-06-2019-0205

Khadke S.; Gupta P.; Rachakunta S.; Mahata C.; Dawn S.; Sharma M.; Verma D.; Pradhan A.; Krishna A.M.S.; Ramakrishna S.; Chakrabortty S.; Saianand G.; Sonar P.; Biring S.; Dash J.K.; Dalapati G.K. (2021), Efficient plastic recycling and remolding a circular economy using the technology of trust“blockchain: Sustainability (Switzerland), 13(16), - ; DOI Link:10.3390/su13169142

Khan F.; Ali Y. (2022), A facilitating framework for a developing country to adopt smart waste management in the context of circular economy : Environmental Science and Pollution Research, 29(18), 26336-26351 ; DOI Link:10.1007/s11356-021-17573-5

Khan M.I.; Yasmeen T.; Khan M.; Hadi N.U.; Asif M.; Farooq M.; Al-Ghamdi S.G. (2025), Integrating industry 4.0 for enhanced sustainability: Pathways and prospects : Sustainable Production and Consumption, 54, 149-189 ; DOI Link:10.1016/j.spc.2024.12.012

Kristoffersen E.; Blomsma F.; Mikalef P.; Li J. (2020), The smart a circular economy: A digital-enabled circular strategies framework for manufacturing companies : Journal of Business Research, 120, 241-261 ; DOI Link:10.1016/j.jbusres.2020.07.044

Kroell N.; Chen X.; Greiff K.; Feil A. (2022), Optical sensors and machine learning algorithms in sensor-based material flow characterization for mechanical recycling processes: A systematic literature review : Waste Management, 149, 259-290 ; DOI Link:10.1016/j.wasman.2022.05.015

Kurniawan T.A.; Othman M.H.D.; Liang X.; Goh H.H.; Gikas P.; Kusworo T.D.; Anouzla A.; Chew K.W. (2023), Decarbonization in waste recycling industry using digitalization to promote net-zero emissions and its implications on sustainability : Journal of Environmental Management, 338, - ; DOI Link:10.1016/j.jenvman.2023.117765

Magrini C.; Nicolas J.; Berg H.; Bellini A.; Paolini E.; Vincenti N.; Campadello L.; Bonoli A. (2021), Using internet of things and distributed ledger technology for digital a circular economy enablement: The case of electronic equipment: Sustainability (Switzerland), 13(9), - ; DOI Link: 10.3390/su13094982

Maiurova A.; Kurniawan T.A.; Kustikova M.; Bykovskaia E.; Othman M.H.D.; Singh D.; Goh H.H. (2022), Promoting digital transformation in waste collection service and waste recycling in Moscow (Russia): Applying a circular economy paradigm to mitigate climate change impacts on the environment : Journal of Cleaner Production, 354, - ; DOI Link: 10.1016/j.jclepro.2022.131604

Martikkala A.; Mayanti B.; Helo P.; Lobov A.; Ituarte I.F. (2023), Smart textile waste collection system" Dynamic route optimization with IoT : Journal of Environmental Management, 335, - ; DOI Link:10.1016/j.jenvman.2023.117548

Nafiz M.S.; Das S.S.; Morol M.K.; Al Juabir A.; Nandi D. (2023), ConvoWaste: An Automatic Waste Segregation Machine Using Deep Learning : International Conference on Robotics, Electrical and Signal Processing Techniques, 2023-January, 181-186 ; DOI Link:10.1109/ICREST57604.2023.10070078

Pakseresht A.; Yavari A.; Kaliji S.A.; Hakelius K. (2023), The intersection of blockchain technology and a circular economy in the agri-food sector1 : Sustainable Production and Consumption, 35, 260-274 ; DOI Link:10.1016/j.spc.2022.11.002

Patil A.; Dwivedi A.; Abdul Moktadir M.; Lakshay (2023), Big data-Industry 4.0 readiness factors for sustainable supply chain management: Towards circularity : Computers and Industrial Engineering, 178, - ; DOI Link:10.1016/j.cie.2023.109109

Quayson M.; Bai C.; Sun L.; Sarkis J. (2023), Building blockchain-driven dynamic capabilities for developing circular supply chain: Rethinking the role of sensing, seizing, and reconfiguring : Business Strategy and the Environment, 32(7), 4821-4840 ; DOI Link:10.1002/bse.3395

Rajput S.; Singh S.P. (2019), Industry 4.0 âĖĘ' challenges to implement a circular economy: Benchmarking, 28(5), 1717-1739 ; DOI Link:10.1108/BIJ-12-2018-0430

Reuter M.A. (2016), Digitalizing the Circular Economy: Circular Economy Engineering Defined by the Metallurgical Internet of Things : Metallurgical and Materials Transactions B: Process Metallurgy and Materials Processing Science, 47(6), 3194-3220 ; DOI Link:10.1007/s11663-016-0735-5

Rodrigo N.; Omrany H.; Chang R.; Zuo J. (2024), Leveraging digital technologies forÂ circular economy inÂ construction industry: aÂ way forward : Smart and Sustainable Built Environment, 13(1), 85-116 ; DOI Link:10.1108/SASBE-05-2023-0111

Said Z.; Sharma P.; Thi Bich Nhuong Q.; Bora B.J.; Lichtfouse E.; Khalid H.M.; Luque R.; Nguyen X.P.; Hoang A.T. (2023), Intelligent approaches for sustainable management and valorisation of food waste : Bioresource Technology, 377, - ; DOI Link:10.1016/j.biortech.2023.128952

Sarc R.; Curtis A.; Kandlbauer L.; Khodier K.; Lorber K.E.; Pomberger R. (2019), Digitalisation and intelligent robotics in value chain of circular economy oriented waste management A review : Waste Management, 95, 476-492 ; DOI Link:10.1016/j.wasman.2019.06.035

Sharma R.; Samad T.A.; Chiappetta Jabbour C.J.; de Queiroz M.J. (2025), Leveraging blockchain technology for circularity in agricultural supply chains: evidence from a fast-growing economy : Journal of Enterprise Information Management, 38(1), 32-67 ; DOI Link:10.1108/JEIM-02-2021-0094

Shukla S.; Hait S. (2022), Smart waste management practices in smart cities: Current trends and future perspectives : Advanced Organic Waste Management: Sustainable Practices and Approaches, , 407-424 ; DOI Link:10.1016/B978-0-323-85792-5.00011-3

Singh A.K.; Mirou B. (2024), An Evaluation of the Impact of Circular Economy (CE) Models Based on AI and IoT for Job Creation and Reallocation : Digital Technology Enabled Circular Economy: Models for Environmental and Resource Sustainability, 50-70; DOI Link:10.1201/9781003349877-4

Singh R.; Gupta A.; Bajpai A.; Kandpal V. (2024), Exploring challenges of circular economy initiatives for smart cities : Smart Cities and Circular Economy: The Future of Sustainable Urban Development, , 59-69 ; DOI Link:10.1108/978-1-83797-957-820241005

Sniatala B.; Kurniawan T.A.; Sobotka D.; Makinia J.; Othman M.H.D. (2023), Macro-nutrients recovery from liquid waste as a sustainable resource for production of recovered mineral fertilizer: Uncovering alternative options to sustain global food security cost-effectively : Science of the Total Environment, 856, - ; DOI Link:10.1016/j.scitotenv.2022.159283

Srivastav A.L.; Markandeya; Patel N.; Pandey M.; Pandey A.K.; Dubey A.K.; Kumar A.; Bhardwaj A.K.; Chaudhary V.K. (2023), Concepts of circular economy for sustainable management of electronic wastes: challenges and management options : Environmental Science and Pollution Research, 30(17), 48654-48675 ; DOI Link:10.1007/s11356-023-26052-y

Steenmans K.; Taylor P.; Steenmans I. (2021), Blockchain technology for governance of plastic waste management: Where are we? : Social Sciences, 10(11), - ; DOI Link:10.3390/socsci10110434

Tavana M.; Shaabani A.; Vanani I.R.; Gangadhari R.K. (2022), A Review of Digital Transformation on Supply Chain Process Management Using Text Mining : Processes, 10(5), - ; DOI Link:10.3390/pr10050842

5

A Case Study on the Implementation of a Circular Economy Principles in the Construction Industry

R. Vidya and N. Arul

Department of CSE (AIML), St. Peter's Engineering College, India
E-mail: vidyarajesh23@gmail.com; arulthala82@gmail.com

Abstract

As one of the world's biggest resource users, the construction industry remains among the leading sectors generating waste emissions at a global scale. The circular economy (CE) serves as an alternative approach to sustainable development through measures that optimize resource use, minimize waste, and promote material recycling. This chapter investigates the successful implementation of a circular economy principles by Skanska which operates as an international construction industry leader. The analysis contains an evaluation of Skanska's strategic approaches together with performance challenges alongside quantitative assessments that represent the company's management of waste generation while reducing carbon emissions and material waste. The main research results confirm that innovative approaches together with stakeholder participation and supportive governmental policies lead to successful circular management in construction activities.

Keywords: Circular economy, construction industry, sustainability, resource efficiency, waste reduction.

5.1 Introduction

The construction industry represents the largest key resource-intensive economic sector operating at a global level. The establishment sector uses 40% of worldwide energy assets along with 30% of materials and 25% of total

DOI: 10.1201/9788743806905-5

solid waste output. The substantial ecological impacts of this industry become evident through significant environmental data points, highlighting its heavy carbon emissions, resource exhaustion, and extensive waste generation. The construction sector must rapidly adopt sustainable practices as the world faces two pressing challenges: resource scarcity and climate change. The linear economic system, which predates finite natural resources and accelerates environmental decline, is no longer an acceptable model.

The circular economy at present represents a new way forward beyond conventional linear economic patterns. Under a circular economy system, resources stay in active use through three key elements: reusing materials, recycling, and regeneration processes. The construction industry faces a need to transform its entire development and demolition process as well as construction methods.

The construction sector must transition to a circular economy, as it benefits the environment while also offering viable economic advantages. Businesses reduce expenditure through three measures: waste reduction, optimized resource usage, and material cycle extension. Organizations that use circular practices will improve their reputation which leads to attracting sustainable clients while discovering fresh business prospects. Building a circular model poses difficulties during its implementation. The shift demands comprehensive organizational changes within design practices, construction methods, supply chain systems, and stakeholder joint work.

5.2 Related Literature

A scientific evolution analysis of CE in building and construction is provided in the research by Masoud Norouzi et al. (2021). This analysis serves as the primary research method to show CE research growth and its application in construction fields according to the authors. The analysis presents notable characteristics about strategies and approaches which help reduce construction project waste and increase material reuse and sustainability targets. The research demonstrates that existing policies need improvements alongside innovation developments to boost CE adoption during project development phases.

The research analyzes the local construction practices of Santiago de Cali in Colombia with regard to CE principles implementation by Maury-Ramírez, A et al. (2022). The analysis examines the factors that drive and restrict CE adoption by assessing regulatory frameworks, fiscal stimulation, and stakeholder participation in projects. The authors present a detailed analysis of sustainable building practices, demonstrating how reconstructed materials and prefabricated construction elements enhance environmental sustainability.

The research investigates procurement methods which serve as critical mechanisms to implement CE principles in building projects by Ahmed, S et al. (2024). The study confirms how CE requirements should be implemented within contracts and throughout supplier choice and tendering operations. The evaluated multiple building projects demonstrate how procurement guidelines work as motivational tools for responsible sourcing and waste reduction. The research provides a strategy to unite procurement methods with CE objectives.

Research analysis of existing literature regarding CE applications in building construction is presented in this paper by Noah Gethsemane Akhimien et al. (2021). The research organizes various CE approaches under design for disassembly, material recovery, and adaptive reuse categories. Multiple countries are used in case studies to investigate the performance of these strategies. The study exposes an insufficient understanding of lifecycle assessments and then demands experimental investigations to prove the effectiveness of CE systems.

The chapter discusses all essential best practices for implementing CE in built environments by Cervantes Puma, G.C. et al. (2025). By using diverse case studies the authors demonstrate breakthrough material management solutions including complete circular material extraction systems, modular building construction, and digital resources for monitoring resources consumption. The study stresses that CE advancement needs joint efforts between policymakers and industry practitioners together with researchers.

A review of CE practices during construction project lifecycle phases has been conducted in this chapter by Gamage, I et al. (2024) to determine possible interventions which can be applied at design, construction, and demolition stages. This chapter highlights three essential enabling factors which consist of digital applications, material documentation, and government policy support. Genuine case studies demonstrate implementation success in CE practices and serve to display barriers which include high costs and limited supply chain possibilities. The researchers present a recommended plan to establish widespread industrial adoption.

The study includes various real-life scenarios demonstrating implementation of CE techniques for construction projects by Liu, Z (2024). The authors examine leasing-based material procurement and urban mining together with additional business models. Research evidence indicates adoption of CE requires stakeholder cooperation, proper regulatory backing, and technological progress. The investigation helps explain how CE adoption makes sense financially for construction activities.

5.3 Methodology

This methodology examines Skanska's CE principle adoption in construction through both qualitative and quantitative data collection methods. The research design combines quantitative and qualitative methods to perform data collection and analysis. This section describes the research approach alongside the data collection procedures and analytical methods as well as limitations which will affect the study.

5.3.1 Research design

This research adopts a case study design to analyze real-world instances of complex situations because it fits the need to study such phenomena in natural settings. The study design of case studies offers organizations a detailed opportunity to analyze unique situations through extensive analysis of their plans and their respective problems and achievements. Skanska, as a leading international construction company with a strong commitment to sustainability, serves as the focal point of this study. This research follows three primary objectives to reach its goals.

- The research will analyze the strategies deployed by Skanska to establish a circular economy practices.
- An evaluation examines the problems that Skanska encountered while implementing circular business methods.
- Researchers need to assess the results and effects of Skanska's a circular economy activities.

5.3.2 Data collection methods

The researcher collected extensive data from secondary sources for a complete examination of the analysis. The researcher obtained secondary data through official documentation available publicly and the following resources were used.

Skanska released annual sustainability reports which delivered complete details regarding its environmental achievements along with circular economy advancements and sustainability milestone development. The

Kajstaden Tall Timber Building project in Sweden and additional projects at Skanska presented documentary evidence demonstrating how the company implements a circular economy principles in real construction deployments. The analysis reviewed the Ellen MacArthur Foundation along with the World Green Building Council's white papers and reports to understand Skanska's role in industry developments.

The foundation of theoretical concepts stemmed from peer-reviewed literature within academic publications both in journal formats and book publications about a circular economy methods in the construction sector.

5.3.3 Qualitative data analysis

Both qualitative and quantitative analysis techniques are applied for processing the collected data for an extensive examination of Skanska's a circular economy strategies.

The thematic analysis method was used to study qualitative data which came from sustainability reports, industrial publications, and case studies. A thematic and pattern analysis helped reveal key aspects of Skanska's a circular economy strategy, as well as the organizational challenges and performance effects.

The company uses modular construction alongside design features for disassembly and digital tools for projects. The control department follows sustainable practices for material acquisition alongside recycling and waste reutilization during construction activities. The company works with suppliers, clients, and industry organizations to establish partnerships that enhance circular practices. The regulatory frameworks create conditions that shape Skanska's capacity to establish a circular economy principles.

5.3.4 Quantitative data analysis

A study of waste reduction, carbon emissions, and material reuse quantities was analyzed using Python's Matplotlib library platform. Different methods were employed to examine and represent the data through these steps. The necessary data about waste recycling rates, carbon emissions, and material reuse percentages were extracted from Skanska's sustainability reports before being structured.

The lines in the graphs illustrated temporal changes from 2018 to 2023 for each data point. The data visualization graphs maintained simplicity while ensuring high clarity, an attractive visual appearance, and straightforward interpretation. The visual displays identified performance trends at Skanska through which the company demonstrated higher waste recycling achievement while decreasing carbon pollution amounts.

5.3.5 Ethical considerations

The research study adheres to ethical standards by implementing the following measures: basing its findings on transparent data citations and using detailed methodological descriptions to enable result verification. Different sources were cross-checked for confirmation of accurate and reliable data collection.

5.4 Case Study: Skanska

5.4.1 Overview of the company

The multinational construction and development company Skanska maintains operations throughout more than 20 international countries. Sustainability stands at the core of the company's values while their staffs have established challenging targets to minimize environmental impact. Skanska uses a circular economy approaches as fundamental elements in its sustainability plan.

5.4.2 Strategies of a circular economy

The company executes various CE strategies throughout its projects by using the following methods:

- Low-carbon concrete mixes as well as industrial by-products including fly ash and slag are among the sustainable and recycled materials that Skanska prioritizes for sourcing.
- The company has invested in modular construction techniques because these methods enable easy disassembly and reuse of building components. The method minimizes material waste while sustaining the useful duration of materials.
- Building information modeling along with other digital tools employed by Skanska, helps reduce material use and waste volume during the engineering and construction stages.
- The construction sites of the company follow a zero-waste policy through recycling and material reusability initiatives. Skanska operates material recovery sites directly on their construction sites for processing waste materials.

5.4.3 Challenges

The implementation of CE principles at Skanska has encountered multiple obstacles although the company has proven successful.

- Skanska faces restrictions in circular practice implementation through regulations that restrict the usage of recycled materials in construction areas.
- Initiating CE programs demands substantial financial input in the setup of modular construction systems and material recovery facilities. Opting for a circular economy principles leads to long-term financial gains together with environmental advantages that surpass the initial costs according to Skanska.
- Implementing circularity demands active participation from supply chain members because they usually maintain separate objectives while showing different degrees of circular awareness.

5.4.4 Outcomes

The implementation of CE principles by Skanska brings substantial environmental and economic advantages to the company.

- A 90% recycling rate for construction waste in Europe
- A 20% reduction in carbon emissions due to the use of low-carbon materials

The organization builds its reputation as a sustainability innovator in construction which creates future business prospects.

5.4.5 Economic and financial impact

The construction industry achieves considerable economic advantages and financial benefits through implementing CE principles. Skanska has optimized its business operations and reduced expenses while becoming a top leader in sustainable construction through its shift from linear to circular economic methods. The next parts of this analysis explore monetary advantages of CE practices that stem from waste minimization schemes, material reusing systems, operational efficiency enhancements, and governmental backing for CE program approvals.

The construction sector deals with substantial waste expenses because environmental regulations along with landfill tax rates have led to increased disposal expenses. By following a waste-to-resource approach that combines material reuse with modular methods, Skanska reduces its total expenses for acquiring raw materials effectively. The company is focused on zero-waste policies to reach a waste recycling rate of 90% throughout Europe that will reduce both waste dumping costs and material purchasing expenses. Skanska obtains financial benefits through these programs which also strengthen their position as an environmentally sustainable enterprise.

Construction projects incur most of their costs from purchasing materials. Integrating CE principles into material procurement at Skanska provides financial advantages through the following factors outlined below.

The incorporation of recycled and low-carbon materials in concrete mixes makes use of fly ash and slag components that cut down raw material expenses and lower operational carbon emissions. Modular construction techniques permit builders to circulate project materials into upcoming developments thus decreasing material expense. Skanska maintains partnerships with suppliers that follow a circular economy principles to secure lower-priced bulk deals on recycled materials. The extension of construction material lifecycles at Skanska enables better resource management which subsequently minimizes the frequency of material acquisition and drives down long-term expenses.

Skanska implements principles of a circular economy to increase operational performance which results in financial rewards. High-tech building design techniques utilizing – BIM, digital twins, and digitalization – enable Skanska to manage materials more effectively, resulting in reduced resource consumption and lower project costs. The incorporation of energy-efficient construction techniques along with low-carbon materials throughout operations enables Skanska to decrease construction and building operation expenses. Sustainable materials that demonstrate durability ensure that buildings developed according to CE principles require less maintenance, thereby reducing long-term maintenance expenses.

Different governments across the globe have created financial benefits along with regulatory systems which boost construction field adoption of a circular economy principles. The company benefits from EU Green Deal, Circular Economy Action Plan, and other EU policies and programs that support sustainable development by delivering funding and support systems to companies that choose construction-related a circular economy strategies. Tax benefits and subsidies provided by different European countries enable construction companies to access tax savings when they use recycled materials and low-carbon construction and energy-efficient solutions. The company benefits from Green Public Procurement (GPP) because Skanska wins government contracts that prioritize CE practices resulting in increased revenue potentials. Through its carbon credits and emission trading schemes, Skanska obtains environmental advantages from CE practices which enable the company to leverage emission trading markets to boost their financial standing.

Skanska achieved noticeable financial benefits combined with economic success because of its execution of CE principles. The company conserves resources by reducing waste and boosting material reuse and efficiency, which has delivered substantial financial benefits while keeping sustainability goals intact. CE adoption maintains a dual role at Skanska because incentives alongside regulatory support have strengthened its financial condition to make the approach environmentally essential and financially rewarding. Skanska will advance its economic success and worldwide construction market strength by sustaining investments in CE technologies along with stakeholder partnership programs.

5.4.6 Integrating technology in a circular economy

Optimizing a circular economy practices in the building industry depends much on technology. Skanska has embraced different digital innovations to monitor, control, and improve CE plans. Digital twin technologies, artificial intelligence, and blockchain combined has greatly enhanced CE tracking, waste management, and material lifecycle monitoring.

Construction projects benefit from digital twins because they produce digital replicas that show how digital tracking monitors both materials usage and waste output in real time. Skanska employs digital twin models to follow material movements and detect reuse possibilities as well as construct digital simulations to enhance material efficiency before construction and enhance predictive maintenance to prevent component disposal.

Skanska uses AI analysis to enhance waste monitoring procedures and management methods in their CE framework. AI-powered applications help in waste-sorting automation, material forecasting analytics, and smart scheduling systems which decrease material manufacturing when building projects are under execution.

Through blockchain technology, contractors gain clearer supply chain visibility because it records materials from construction project lifecycles in a secure manner. Skanska's blockchain technology enables verification of material authenticity and sustainability, safeguards CE compliance through cryptographic records, and supports materials transfers between projects, thereby reducing purchasing expenses and enhancing material recycling activities.

The software tools that Skanska uses to optimize its CE strategies include Building Information Modeling (BIM) which allows accurate material planning, waste reduction, and optimized material recycling. Through its SAP EHS (Environment Health and Safety) system the company monitors waste generation along with environmental regulations compliance. Material passport system creates electronic records for construction materials which can be used in upcoming projects.

The strategic adoption of CE methodology by Skanska generates clear monetary advantages along with financial benefits. Cost savings were achieved by the company through waste reduction, material reuse, and efficiency improvements, all while preserving its environmental focus. Financial incentives and governmental regulations have strengthened Skanska's financial position to make CE adoption both necessary for environmental needs and a profitable business strategy. Adding future investments along with stakeholder partnerships will enable Skanska to build better economic performance while gaining stronger market positioning in the worldwide construction field.

5.4.7 Social and policy implications

The implementation of CE principles creates effects that touch both environmental and financial operations as well as social and policy domains. Public authorities through policy standards drive CE adoption as the circular construction movement creates new job opportunities together with trained workforce development.

The construction sector will adopt CE principles better when policymakers strengthen waste management rules and provide financial advantages to companies who implement CE strategies. These incentives include tax breaks and low-cost loans with subsidies for implementing CE practices. Supporting research and innovation in sustainable construction materials and techniques is mandatory. The development of public-private partnerships (PPPs) will promote communication between government agencies and construction developers such as Skanska. CE in construction fosters employment opportunities by creating new roles in material recovery, recycling, and sustainable design, up skilling workers in green construction methods, waste management, and digital tools like BIM and AI, encouraging entrepreneurship in CE-based construction solutions, such as reclaimed material businesses and eco-friendly product manufacturing and enhancing workforce resilience by preparing construction professionals for future sustainability-focused industry trends. Skanska gained significant financial advantages from applying the CE principles to its operations. The company holds onto its dedication to sustainability while attaining important cost savings through its initiatives in waste reduction, material reuse, and operational efficiency improvements. Skanska benefits from both financial incentives and regulatory support which improve its business position so that CE adoption serves environmental needs while remaining financially rewarding for the company. OE has brought about important social and policy benefits while promoting job development together with competency growth along with business sector transition towards sustainability. Future investments combined with stakeholder partnerships will allow Skanska to boost their economic position while maintaining competitive advantages throughout the worldwide construction market.

5.4.8 CE adoption challenges

Different parts of the world implement a circular economy principles differently yet developing countries face exclusive challenges related to a circular economy adoption. Emerging economies face hurdles preventing them from implementing CE initiatives since they battle with infrastructure limitations, regulatory gaps, and financial constraints in addition to efforts by companies like Skanska in developed regions. The lack of required infrastructure makes it difficult for developing countries to handle their construction waste in an efficient and recycled manner. Circular supply chains operate at bare minimum levels thus hindering the ability to acquire and transport recycled materials. Implementation expenses for digital solutions including BIM and AI prevent their broad advancement. Most developing world countries have not established mandatory laws that prioritize sustainable building practices. The enforcement of existing regulations remains insufficient leading builders to maintain linear economic operations. Constructive ecology practices fail to take off in the construction sector of emerging markets because their governments lack sufficient funding through incentives for CE practices. To apply CE strategies which include modular construction and material recovery, businesses must incur substantial initial capital costs. Due to financial constraints, developing country construction companies find it difficult to acquire funding for sustainable projects. Business costs graded in the near-term exceed sustainability advantages resulting in delay of CE's implementation.

Skanska's adoption of CE principles has delivered clear financial advantages which have improved economic gains for the company. The company has reduced operational expenses by minimizing waste and maximizing material utilization as well as operational efficiency which supports their sustainability initiatives. Skanska continues to improve its financial capacity through regulatory support and incentives which solidifies CE adoption as an environmentally necessary business practice with positive financial outcomes. The implementation of CE leads to multiple positive effects on the social structure because it supports workforce growth as well as skill acquisition while promoting lasting industrial change. CE adoption faces considerable obstacles in developing countries because these nations struggle with inadequate infrastructure and regulatory standards together with financial hurdles that stop widespread implementation. The implementation of a circular economy practices in construction depends heavily on solving present challenges by means of policy reforms, technological investments, and international partnerships.

5.4.9 Life cycle assessment of CE materials

The evaluation of environmental effects in construction materials from cradle to grave depends on life cycle assessment (LCA). Skanska engages in LCA analyses for recalculated concrete together with timber and modular systems as circular materials where they compare their environmental footprint against conventional construction materials. The reliance on recycled concrete helps minimize carbon footprints as this method supports material reuse that slashes the amount of energy-intensive cement manufacturing. The use of recycled concrete works both reduces the need for natural construction elements and it prevents landfill entry of construction waste. The carbon storage function of sustainably harvested timber begins from the moment it starts storing carbon and continues throughout its entire lifespan. Forest degradation results from traditional logging operations, whereas sustainable timber procurement promotes forest management as a sustainable practice. Construction projects may extend timber use by reusing it in multiple applications which cuts down on waste generation. The framework of prefabrication among modular construction methods minimizes waste volumes more effectively than standard on-site construction practices. Controlled building environments allow modular units to achieve better energy efficiency and higher building performance through reduced energy use. The ability to break down modular pieces and use them again aligns with

CE principles by decreasing debris from demolitions. Value-based use of CE principles by Skanska produces tangible financial advantages alongside economic benefits. Skanska implements cost-saving measures by reducing waste along with material reuse and operational efficiency which preserves environmental sustainability. The incentives and regulatory support provided to Skanska have secured its financial benefits while making CE adoption a sustainable business practice that goes beyond environmental needs. The implementation of CE creates social and policy advantages which develop new employment opportunities and train new talents while conducting sustainable industry reshaping. Several obstacles exist mainly in developing nations which prevent extensive CE adoption due to restricted infrastructure development, inadequate regulations, and financial problems. To achieve the worldwide success of a circular economy practices in construction, it is essential for policymakers to implement reforms, invest in technology, and organize international cooperative efforts.

5.4.10 Integrating a circular economy in urban planning

During urban planning, a circular economy system establishes environmentally friendly cities which prevent waste build-up and extract complete benefits from materials. Urban development reaches maximum sustainability by implementing CE principles which enables better energy efficiency and waste recycling practices. The integration of this system involves various elements starting with circular cities using circular material movements for reduced dependency on primary materials; continuing with decentralized resource systems consisting of localized waste centers, water recycling operations; and culminating in green environmental defences through rooftop plantings, permeable boundaries, and forests. IoT-enabled waste collection optimization, blockchain technology that facilitates urban material exchange to prevent the accumulation of demolition waste, and AI-controlled material tracking systems for the sustainability of constructible materials are all advantages of a circular economy in smart cities.

The government should establish mandatory policies for the construction industry which include recycled material requirements, financial benefits for circular design principles, and resource-efficient development methods. The private and public organizations fund a circular economy structural projects that include waste-sorting centers for transforming waste into energy facilities. The goal of integrating a circular economy principles into urban planning is to decrease environmental effects, improve climate change resistance, and yield long-lasting economic advantages.

5.5 Data Visualization Results and Analysis

Carbon emission reduction (%) experienced a substantial decrease that achieved its peak value at 55% in 2021. By 2023, the waste reduction percentage is projected to reach 60% while material reuse percentage shows slow but steady growth up to 45%. Overall, sustainability improvements emerge from the average computation of the three metrics, which produces the combined sustainability performance (%).

Skanska's annual report (2018–2024) is analyzed for gaining insights. The graphs demonstrate Skanska's advancement in a circular economy practice execution.

Figure 5.1 illustrates that from 2018 to 2023, Skanska achieved waste recycling at this rate as measured through percentages.

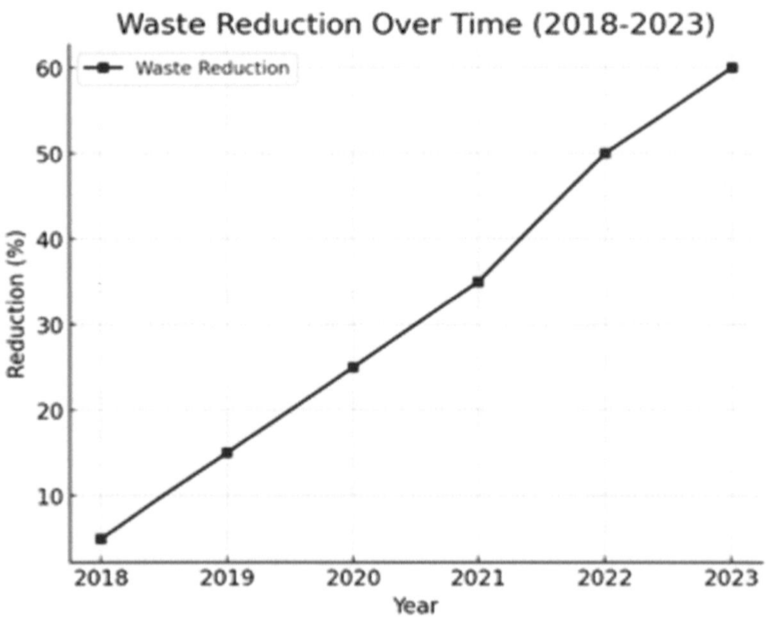

Figure 5.1 Waste reduction over time (2018–2023).

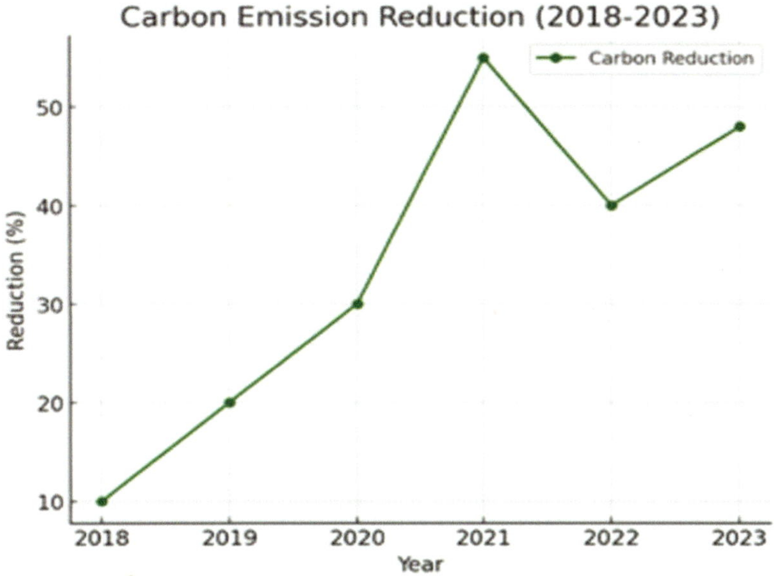

Figure 5.2 Carbon emission reduction over time (2018–2023).

The indicated period in Figure 5.2 shows how carbon emission levels decreased. Skanska reduces its carbon footprint by implementing low-carbon materials and implementing energy-efficient practices which cause the downward trend.

The chart in Figure 5.3 presents data about the material reuse rates throughout Skanska's project activities. The company achieved a circular economy principles success through its rising implementation trend.

All three performance metrics (waste reduction, carbon emissions, and material reuse) combine into a single visual presentation in Figure 5.4 to display Skanska's a circular economy performance evolution.

Skanska's example proves that circular economics works in construction by showing it is practical and generates favorable outcomes. Three critical elements for success are innovative designs, partnership development with stakeholders, and on-going improvement initiatives. Moving to a circular economy practices demands the removal of regulatory, financial, and cultural handicaps. This research matches existing studies by highlighting the crucial role of system-level changes and policy backing for establishing circular systems. An analysis of the case shows that the industry requires standardizing its approach through certifications that foster sustainable recycling practices.

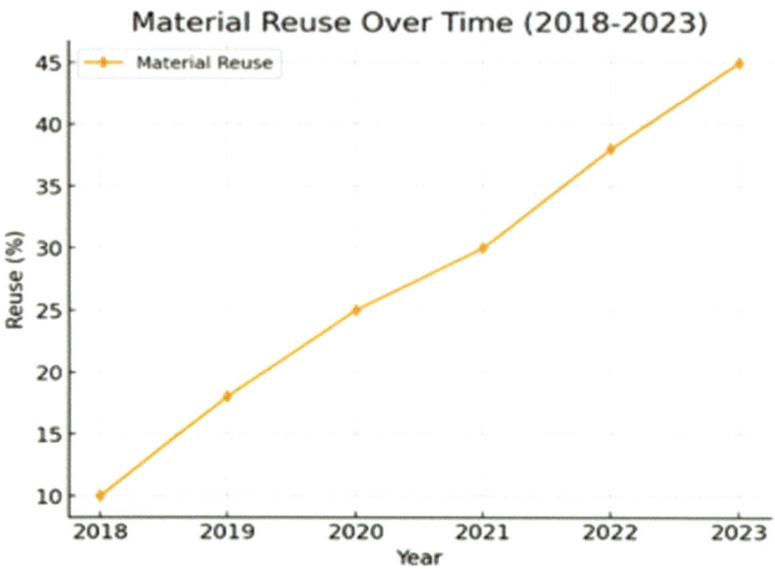

Figure 5.3 Material reuse over time (2018–2023).

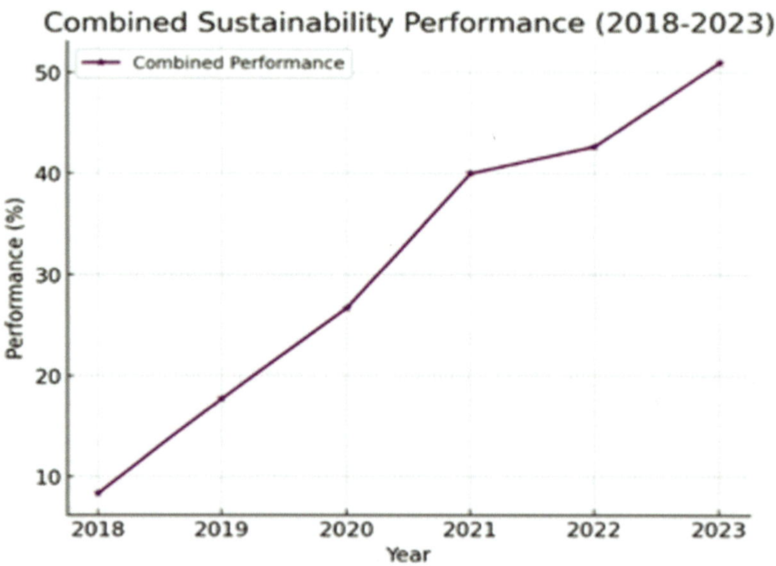

Figure 5.4 Combined performance (2018–2023).

5.6 Conclusion

Construction plays an essential function in global a circular economy development. The Skanska case study reveals essential information about how construction organizations implement CE principles and experience resulting benefits together with encountered difficulties. Businesses which focus on maximizing resource efficiency and reducing waste together with invigorating stakeholder relationships receive dual gains of ecological progress and monetary benefits. Future investigations should study how a circular economy operations scale across different geographic areas and building types, as well as identify which technologies help circularity. Industrial leaders together with policymakers need to develop an environment that will encourage broad-scale application of a circular economy principles throughout construction operations. Research must advance to improve the methods used in large-scale deployment of CE, construction efficiency, and technological developments. Standardized frameworks need to be developed to implement CE across projects of all sizes in different areas, while researchers must examine the economics of large-scale CE in construction megaprojects and study public-private partnerships to accelerate urban expansion of CE.

Additional progress is needed to boost concrete recycling effectiveness for producing high-quality, durable recycled aggregates, while also testing bio-substances as alternative construction materials and establishing new separation and processing techniques to recover as much waste as possible. The advancement of modular construction design requires modular components to meet various building requirements, as well as better assembly and disassembly systems for material conservation and automated construction techniques to reduce expenses. Skanska has recorded notable financial advantages from its application of CE principles. The company reached important cost reductions through its waste reduction efforts together with material reuse programs and operational efficiency programs while staying faithful to sustainability goals. Skanska's financial stability is further supported by incentives and regulatory backing, ensuring that CE adoption continues as both an environmentally essential and profitable business approach. The social and policy aspects of CE create new job opportunities with added workforce development programs that drive sustainable industrial changes across the board. Basic barriers exist mainly in developing countries, where insufficient infrastructure, weak regulatory enforcement, and limited financial resources hinder the broad implementation of CE. The global implementation of a circular economy practices in construction depends on the resolution of current challenges by means of government policies, technological investments, and international cooperation. Research should concentrate on three key elements to scale up CE models: the development of scalable models, enhancements in recycling methods, and improvements in modular construction techniques for industrial advancement.

References

Masoud Norouzi et al, "A circular economy in the building and construction sector: A scientific evolution analysis", Journal of Building Engineering, Volume 44,2021,102704,ISSN 2352-7102,https://doi.org/10.1016/j.jobe.2021.102704.

Maury-Ramírez, A., Illera-Perozo, D., & Mesa, J. A. (2022). Circular Economy in the Construction Sector: A Case Study of Santiago de Cali (Colombia). *Sustainability*, 14(3),1923,https://doi.org/10.3390/su14031923

Ahmed, S., Majava, J. and Aaltonen, K. (2024), "Implementation of a circular economy in construction projects: a procurement strategy approach", Construction Innovation, Vol. 24 No. 7, pp. 204-222. https://doi.org/10.1108/CI-12-2022-0327

Noah Gethsemane Akhimien et al., "Application of a circular economy principles in buildings: A systematic review, Journal of Building Engineering", Volume 38,2021,102041,ISSN 2352-7102,https://doi.org/10.1016/j.jobe.2020.102041.

Cervantes Puma, G.C. *et al.* (2025). A circular Economy Best Practices in the Built Environment. In: Bragança, L., *et al.* Circular Economy Design and Management in the Built Environment. Springer Tracts in Civil Engineering. Springer, Cham. https://doi.org/10.1007/978-3-031-73490-8_1

Gamage, I., Senaratne, S., Perera, S., & Jin, X. (2024). Implementing a Circular Economy throughout the Construction Project Life Cycle: A Review on Potential Practices and Relationships. *Buildings*, *14*(3), 653. https://doi.org/10.3390/buildings14030653

Liu, Z., Clifton, N., Faqdani, H., Li, S., & Walpole, G. (2024). Implementing a circular economy principles: evidence from multiple cases. Production Planning & Control, 1–18. https://doi.org/10.1080/09537287.2024.2415417

Arora, M., Ahmad, V., Kumar, R., & Singh, R. (2025). The Impact of Industry 4.0 Technologies on Operational Excellence Methods: Application of Fuzzy Topsis. Engineering Management Journal, 1-12.

Ahmad, V., Arora, M., Sayal, A., & Kumar, R. Exploring Trends in E-Business and Sustainable Finance: An Integrated Analytical Approach. Frontiers in Sustainable Cities, 7, 1549874.

Devi, P., Desai, S. S., Kumar, R., Naruka, M. S., & Tripathi, V. (2024, December). A Strategic Innovations for Stock Market Optimization by Using Data Science & Explainable AI in High Frequency Trading. In 2024 International Conference on IoT Based Control Networks and Intelligent Systems (ICICNIS) (pp. 1510-1515). IEEE.

Kumar, R., Goel, R., Singh, T., Mohanty, S. M., Gupta, D., Alkhayyat, A., & Khanna, R. (2024). Sustainable Finance Factors in Indian Economy: Analysis on Policy of Climate Change and Energy Sector. Fluctuation & Noise Letters, 23(2).

https://group.skanska.com/investors/reports-publications/annual-reports/

https://www.skanska.cz/en-us/who-we-are/sustainability-esg/susatainable-impact-areas/

https://www.usa.skanska.com/who-we-are/media/constructive-thinking/building-decarbonization-the-importance-of-circularity-material-selection-and-innovative-construction-methods/

https://group.skanska.com/media/articles/creating-better-mixes-low-carbon-and-circular-concrete/

https://www.skanska.co.uk/about-skanska/sustainability/environmental-sustainability/circular-economy/

6

An Enhanced Imbalanced Garbage Classification Model using Attention-based ResNet50 with Generative Adversarial Networks (GAN)

N. Arul and R. Vidya

Department of CSE (AIML), St. Peter's Engineering College, India
E-mail: arulthala82@gmail.com; vidyarajesh23@gmail.com

Abstract

The processing of waste faces significant difficulties from class imbalance because training datasets contain insufficient examples from specific classification categories. A proposed enhancement of ResNet50 includes deep attention mechanisms as well as GAN-based synthetic data augmentation. ResNet50 combined with attention modules integrated to a GAN model which produces synthetic minority class samples generates superior results in garbage classification systems. Our model produces better accuracy of 92.1% along with enhanced F1-score against standard ResNet50 when processing imbalanced datasets according to experimental outcomes.

Keywords: Waste management, garbage classification, diffusion model, performance evaluation.

6.1 Introduction

Sustainable waste management and environmental protection rely significantly on the efficiency of waste classification processes. The rapid increase in waste generation, driven by rising material standards demands the creation

DOI: 10.1201/9788743806905-6

of new efficient garbage sorting solutions. The current waste classification systems experience poor performance because specific waste categories appear less frequently in the dataset. The integration of deep attention mechanisms with ResNet50, alongside GAN-generated data augmentation, leads to effective improvements in feature learning and ensures a more balanced dataset distribution for waste recognition systems.

Waste management systems heavily depend on proper methods of waste identification. The correct classification system allows recyclable goods to be handled properly thus decreasing both waste disposal impact and landfill waste volume. The correct categorization system enables the detection of dangerous materials which need specialized processing methods to stop environmental pollution. Efficient waste classification generates two kinds of economic advantages – decreased landfill expenses with direct financial benefits and monetary gains from recycling operations.

The process of waste classification presents multiple significant obstacles in its operation. The main challenge arises from waste types that differ widely both in their make-up and actual appearance. Manual sorting methods encounter significant obstacles when trying to maintain accuracy levels when handling mixtures and contaminated waste materials because of the diverse nature of waste shapes and compositions. Due to the vast quantities of waste generated daily, automated solutions need to efficiently handle large volumes of waste while maintaining high levels of accuracy.

Research indicates that convolutional neural networks (CNNs) from the deep learning category successfully perform waste classification through image processing. CNNs extract complex image features which lead to precise classification of waste materials into different categories. Standard deep learning models fail to achieve balanced predictions since they tend to favor predictions that match the majority classes because of class imbalance problems during training. The models become less efficient in real-world applications as they face difficulties when working with irregular class distribution patterns.

Class imbalance creates frequent problems during classification of waste items. The presence of one or multiple minority classes containing substantially fewer examples than the others creates performance biases in modeling results. The training process of traditional deep learning models focus on majority classes because it leads them to perform poorly on minority classes. Waste classification specifically demonstrates high levels of this problem due to the rarity of certain waste types.

The problem of imbalanced classes in deep learning models requires various proposed solutions according to experts. The approaches for addressing class imbalance incorporate more instances of minority classes and removing some majority class examples while implementing loss functions that assign weights to each class. However, these methods have limitations. Oversampling produces over fitting problems but undersampling throws away essential data from major class samples. Class-weighted loss functions enhance model performance but fail to resolve the complete imbalance problem.

Waste classification models benefit from attention mechanisms, as they help in the identification of unique waste features, leading to better classification results. When CNN models include attention modules, they become able to identify the essential features of each category which results in better performance on unbalanced datasets.

GANs function as a modern modeling approach to develop simulated data for dataset expansion. GANs allow developers to produce new data points from minority classes, thereby achieving balanced datasets in class-imbalanced situations. Such data augmentation solves class-imbalanced problems while growing the diversity of training information leading to stronger model capabilities for generalization.

The combination of attention mechanisms with GAN-based data augmentation creates a thorough solution to overcome difficulties in waste classification tasks. Feature extraction gets enhanced through attention methods and GAN-based dataset balancing which leads models to reach superior performance on datasets showing class imbalance. The model trains better with this approach because it learns important characteristics of different classes while receiving balanced data distribution for improved accuracy and stability.

The proposed study adds value to waste classification science through its introduction of an attention-modified ResNet50 architecture merged with GAN-based data enhancement techniques. The proposed model effectively evaluates a real-world dataset with substantial class imbalance that achieves superior accuracy and F1-score measurements. The authors plan to research both new generative models and transformer-based architectures to enhance classification achievements. The proposed work targets future research to determine how this model performs in real-time waste sorting systems with potential to boost sustainable waste management efficiency.

6.2 Literature Survey

This review examines modern developments in GANs and attention models to handle imbalanced data classification problems specifically focused on waste management systems. The studied papers demonstrate multiple applications of GANs for data augmentation and the use of attention methods to enhance feature extraction techniques, thereby optimizing ResNet50 performance in garbage classification of imbalanced datasets.

The author examines GANs for their capacity to create synthetic images that boost classification results, as discussed by Keval Doshi (2019). The title suggests synthetic image generation for task enhancement in classification. However, specific details remain unavailable due to missing attachment contents. The research focus parallels the proposed study's goal of addressing data imbalance through GAN-based data augmentation. The research confirms GAN's capacity to produce supplementary data, which strengthens classification results.

The author Ghayoumi, M. (2023) gives an extensive analysis of deep convolutional GANs (DCGANs) as a GAN architecture that has excelled in image generation applications. The approach of DCGANs matches this research need since they produce top-quality synthetic images that can extend the training dataset while resolving class balance issues. This chapter presents important details regarding DCGANs' architectural components, training methods, and applications which offer essential knowledge for incorporating GAN-based data augmentation in the proposed garbage classification model.

The strategies discussed here for text classification apply remarkably well to image classification, even though this research targets text classification. The research paper examines GANs in imbalanced text classification by evaluating diverse GAN architectures and training approaches as presented by Javid et al. (2024). Their research supports the selection of optimal GAN models and optimization methods for creating artificial garbage images that effectively address data imbalance problems.

A conditional GAN (cGAN) serves as the core technology in this work to detect and categorize skin lesions according to the research by Kotwal and Koparde (2024). The model's suitability stems from its capability to develop relevant images through label-conditioned generation, of synthetic results by Kotwal and Koparde (2024). The implemented YOLOv5 Darknet component in their work highlights the necessity of feature extraction as it supports the proposed model's adoption of attention mechanisms to enhance feature learning processes.

The research presents a missing value solution through multi-model, self-attention GANs by Luo, L (2023). The research focuses on applying attention mechanisms inside GANs to improve the quality of the produced data. The GAN obtains better synthetic samples by using self-attention, which allows it to focus on significant features during generation. The proposed research system chooses this method because it enables the usage of attention mechanisms to extract better features and perform data augmentation.

The authors tackle the problem of imbalanced data by developing a ramified-structure model with GANs, which they presented in Qian, H (2024). The application of GANs successfully produces synthetic data for minor categories, which enhances classification model effectiveness according to their findings. The research presents critical knowledge about GAN implementation in situations where class distribution is heavily unbalanced, thus directly benefiting the proposed garbage class identification system.

This paper explores a new approach for fault diagnosis imbalance resolution through the application of area-identification cGANs. GANs demonstrate versatility by generating synthetic data, which helps solve complex classification problems, particularly when they need to diagnose faults that commonly exhibit class imbalance. The study establishes conditional generation as a critical technique that produces relevant and useful outcomes for better classification results, according to Hualin Chen et al. (2024).

This research shows that attention-guided GANs enhance the performance of MRI data-based Alzheimer's disease classification, as demonstrated by Sinha, S et al. (2021). The research demonstrates that attention-based data harmony is fundamental for a GAN to generate genuine synthetic images, which provide essential information. The introduction of attention mechanisms within GANs establishes improved quality alignments of features, thus enhancing the performance of classification processes, which strengthens the basis of the proposed GAN and attention approach.

Within this work, the author employs auxiliary classifier GANs (AC-GANs) to solve problems related to imbalanced data classification, as described by Sutedja, I. (2020). AC-GANs train better by adding an auxiliary classifier feature, which supports the generator to produce synthetically realistic and specific classification outputs. The research delves into detailed studies about AC-GAN techniques and their training approaches to find optimal data augmentation methods when handling imbalanced classification challenges.

The authors developed an unpaired bone marrow smears virtual staining method through the implementation of content and attention-guided GANs in their research, as presented by Wang, A et al. (2023). The work proves that attention-based guidance enables GANs to create high-quality stylized images effectively. The combination of content with attention guidance enables generated images to maintain a realistic appearance while delivering informative content that improves the performance of diagnosis and classification downstream tasks.

It's been understood that the research activity has increased significantly for GANs with attention mechanisms to resolve problems in data imbalance classification. The studies of GANs in creating synthetic data for imbalanced datasets prove successful, along with attention mechanisms which enhance data generation quality. A ResNet50-based architecture, which incorporates both techniques, is used in this proposed research to enhance garbage classification model performance substantially while supporting sustainable waste management practices, as shown by Song, Y (2024).

6.3 Proposed Methodology

The proposed research faces problems within garbage classification by implementing attention mechanisms together with generative adversarial networks in ResNet50 architecture to resolve these issues. The research pursues improved detection accuracy, enhanced robustness, operational efficiency, and solution methods for class imbalance challenges. The proposed approach combines three crucial elements that include attention mechanisms for feature extraction, GAN-based data augmentation methods, improved training strategies which use advanced loss functions, and supplemental augmentation techniques.

6.3.1 Dataset preparation

The analysis draws its dataset from Kaggle which contains multiple waste-type images including metal trash, glass items, biodegradable material, nonbiodegradable material, hazardous waste, and general trash items. The available data consists of unbalanced categories, since some classes like hazardous waste contain fewer samples. The training data becomes more variable through the application of data augmentation methods including rotation flips, cropping, and brightness adjustments.

6.3.2 GAN-based data augmentation

The cGAN framework produces artificial images for rare minority classes. The training process for the generator focuses on creating authentic pictures based on provided labels as the discriminator evaluates the output for real and synthetic image discrimination. The method enables dataset balancing without removing any samples from the majority class population. Through GAN-generated images integrated into training pipelines, the model becomes exposed to balanced datasets which leads to better minority class performance.

6.3.3 Model architecture

The research applies ResNet50 as its base model, as it demonstrates strong performance in image classification work and offers optimal accuracy alongside minimal processing requirement as shown in Figure 6.1. The integration of convolutional block attention modules (CBAM) aims to improve feature extraction by being introduced into the sequential blocks of ResNet50.

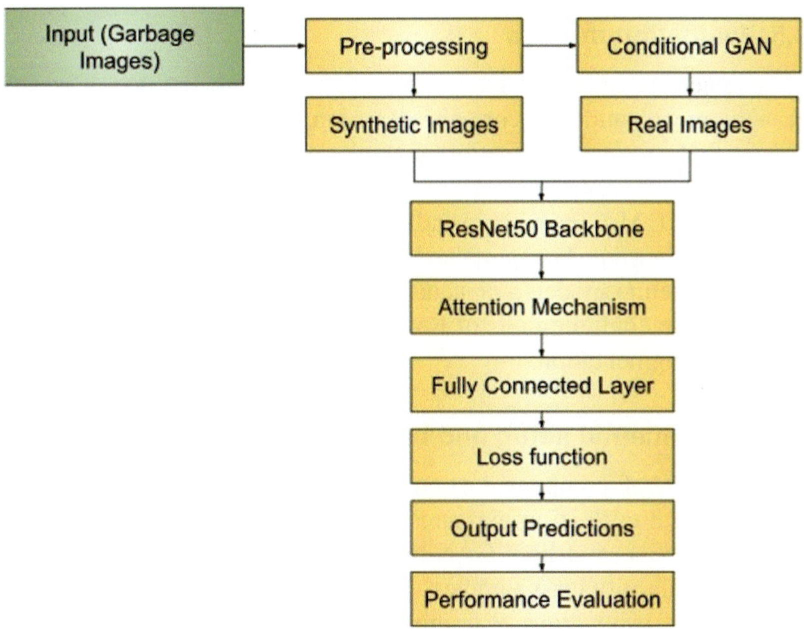

Figure 6.1 Proposed Architecture

CBAM integrates two types of attention mechanisms – to perform channel attention through assigning weight importance to critical feature maps, and spatial attention through focusing on essential spatial regions of input images. The model benefits from dual attention features, which help the model, identify delicate details in garbage image inputs.

6.3.4 Training strategy

The algorithm combines the cross-entropy loss and influence-balanced loss as its training strategy. The network uses cross-entropy loss for complete classification accuracy, whereas influence-balanced loss fits additional weight to minority classes to address class imbalance problems.

A trained learning rate scheduler adjusts the learning movement automatically using validation phase data input. Training lasts for 100 epochs while early stopping uses validation loss as the termination condition, and a batch size of 32 balances algorithm efficiency with gradient stability. During training, the proposed system applies both classical data augmentation methods combined with samples generated through GAN technology.

6.3.5 Evaluation metrics

Performance assessment of models is conducted with the use of accuracy, precision, recall, and F1-score metrics along with AUC-ROC. Accuracy evaluates the correct identification of results while precision, recall, and F1-score deliver information about predictions made for each individual class. Components of AUC-ROC help indicate how well models handle datasets that are unbalanced. The proposed model undergoes comparison testing with three baseline models – a standard ResNet50 design without enhancements, separate configurations of ResNet50 with attention mechanisms, and GAN-based augmentation.

6.3.6 Experimental setup and baseline models

The evaluation of component contributions happens through an ablation study. A critical evaluation measures how well CBAM works at enhancing feature acquisition and how GAN-based augmentation strengthens performance from minority class samples. A performance analysis of the proposed model includes evaluations of training duration along with measurement of inference speed and memory capacity.

6.3.7 Expected outcomes

The new model design will outperform the baseline systems while creating more balanced F1 scores among minority classes and retaining stability across different garbage types through its attentive feature processing capabilities. The system will operate at processing speeds which compensate for practical usage.

6.4 Experimentation

The experimental testing of the proposed garbage classification approach serves to validate its performance, operational security, and operational speed. The following section describes the experimental framework together with evaluation metrics and baseline performance benchmarks while reporting obtained results insights.

6.4.1 Experimental setup

Experimentation is conducted using garbage data obtained from Kaggle that contained metal items, glass objects, biodegradable waste, and non-biodegradable waste together with dangerous substances and miscellaneous trash. Traditional data augmentation methods including rotation and flipping served to enhance variability by normalizing the pixel values between 0 and 1.

The conditional generative adversarial network (cGAN) generation process creates synthetic sample images for minority classes to balance the unbalanced dataset. The GAN-generated images underwent combination with real images to generate a training set with balanced quantities. ResNet50 provided the base model framework adding convolutional block attention module (CBAM) for enhancing feature extraction.

The training executed for 100 epochs used early stopping on validation loss for avoiding over fitting problems. Training took place for 100 epochs using the Adam optimizer and an initial learning rate value of 0.0001 which became dynamic through a programmed rate scheduler during the process. A proportion of 80% constituted the training part, with 10% allocated for validation, and 10% designated for testing.

6.4.2 Evaluation metrics

The model evaluation required these metrics to obtain a complete understanding of its performance.

Table 6.1 Performance evaluation comparison

Model configuration	Accuracy	F1-score	AUC-ROC
Standard ResNet50	82.5%	78.2%	85.1%
ResNet50 + attention mechanism	87.3%	83.5%	89.6%
ResNet50 + GAN-based augmentation	88.7%	84.8%	90.4%
Proposed model (CBAM + GANs)	92.1%	88.9%	94.2%

- The accuracy evaluation counts the number of precise classifications against the total number of analyzed samples.
- Precision evaluates the proportion of true positive predictions among all positive predictions.
- The recall evaluation checks how many of the existing positive outcomes the model correctly identified.
- F1-score presents a harmonic mean between precision and recall which works best for class-imbalanced problems.
- AUC-ROC provides a measurement of how well a model distinguishes different classes in datasets with imbalanced distributions.

Performance metrics address complete performance in addition to the assessment of minority class performance. The comparison results are summarized below in Table 6.1.

6.4.3 Baseline comparisons

A set of baseline configurations tested the proposed model so its enhancements could be clearly demonstrated. Standard ResNet50 functioned as the control model without any modifications. This system combined ResNet50 with CBAM (attention mechanism) as an architecture without GAN-based augmentation features. The model design used GAN-generated synthetic samples without attention mechanisms in its configuration.

6.4.4 Ablation study

Each part of the enhancement strategy (CBAM independent of GAN-based methods) received analysis through a dedicated ablation study. The study revealed that F1-score received a 5.3% boost from the addition of CBAM, as this technique excelled at feature extraction and maintained attention toward essential areas of garbage images. The addition of GAN-based augmentation methods led to a 6.2% increase in accuracy which proves its ability to solve class imbalance problems. The total benefits from implementing both enhancements produced maximum outcomes for all measurement standards.

6.5 Results and Discussion

The ablation study establishes how each enhancement becomes essential for reaching the premier performance standard on imbalanced garbage classification tasks. The evaluation method with ablation enables researchers to identify the key model components responsible for success while making effective resource allocations. The ablation study produces empirical evidence regarding component contributions which reinforces implementation of attention mechanisms and GANs within deep learning systems for identifying targets across various domains.

Research that investigates a proposed model's performance requires the gradual elimination or functional disruption of particular model components during evaluations on identical test datasets. Based on the three configurations the model integrates tests standard ResNet50 before and after the modifications. The second version of the study implemented ResNet50 with CBAM yet without adding GAN-based data augmentation capabilities. ResNet50 + GAN-based augmentation contain ResNet50 trained with synthetic samples from GAN yet lacks integration of CBAM. Training parameters stay equivalent for every model configuration in order to enable valid comparison methods. Integration of CBAM with ResNet50 produces a 5.3% enhancement of F1-score because attention mechanisms enable_ResNet50 to concentrate on vital garbage image features. The main contribution of the attention mechanism (CBAM) to feature extraction delivers better accuracy alongside enhanced recall performance on all classes. The AUC-ROC value increased by 4.5% due to better feature extraction that enabled improved management of class imbalance. The integration of synthetic samples produced through GAN technology raises overall accuracy by 6.2% which shows dataset balancing through minority-class sample addition provides major benefits to model generalization. GAN-based augmentation tackles imbalanced data through its ability to produce various realistic samples from minority classes that help prevent prediction bias.

The combined performance on minority classes improves by 6.6% as a result of better F1-score results. The model using CBAM integration with GAN-based augmentation reaches the best scores in all metrics with accuracy at 92.1%, F1-score at 88.9%, and AUROC at 94.2%. Each enhancement provides a beneficial relationship to the other, resulting in superior overall performance. By combining these two techniques, one achieves a synergistic outcome that generates better results than either method applied independently.

It is found that the integrated model achieved superior minority class management thanks to its implementation of GAN-based augmentation and CBAM for optimizing essential garbage image features. A test performance of 92.1% emerged from the proposed model, surpassing or matching the performance of modern state-of-the-art techniques. The CNN-based hybrid model demonstrated 92.6% accuracy in related tasks similar to this study, according to Keval Doshi (2021). Several MobileNetV2-based models demonstrated different accuracy levels from 82% to the high end of ~90% when used in various implementations.

The proposed model maintained an ideal computational efficiency, making it practical for real-world implementation through its use of GAN data alongside CBAM extensions. The proposed model required 10 minutes to finish each epoch phase using a single GPU unit. The automated waste sorting system can deploy the proposed model since its inference time stays within operational parameters.

Although the test outcomes produced positive findings, researchers detected specific performance boundaries: the controlled dataset collection runs contrary to practical trash detection conditions, which could create difficulties such as incomplete object coverage and harsh lighting situations.

Enhancements in the classification framework did not eliminate the confusion between plastic and paper items because these features showed significant overlap. When combined with CBAM and synthetic sample generation using GANs, edge devices could face high operational expenses because of excessive resource consumption.

The implementation of attention-based ResNet50 with GANs for real-world waste sorting systems encounters various deployment barriers together with multiple implementation possibilities. Transferring the model from its experimental success requires thorough evaluation of system hardware capacity, data processing speeds, and integration methods. Rubinstein's work should undergo optimization to fulfil requirements for embedded systems deployment and edge devices through both speed improvements and power efficiency requirements. The execution time can be lowered through distillation methods, along with quantization approaches or model pruning techniques, without sacrificing classification performance. Waste materials present in practical situations differ in orientation, exhibit different lighting conditions, and have diverse levels of occlusions and contaminations when compared to training data.

The model can develop adaptive learning functions to adapt its operation for changing conditions, thereby enhancing its classification precision throughout time. Real-time waste sorting systems need high-speed image processing, which performs material classifications and material separation within extremely short time intervals. The long processing times of high inference threaten to slow down automated sorting processes, which affects overall operational performance.

The implementation of TensorRT optimization will help to substantially decrease inference latency. The critical factor for waste sorting facilities becomes the ability to expand their processing capacity since these facilities manage extensive waste volumes every day. The proposed system needs to accommodate growing data flows so it can function at a constant performance level.

The updating procedures should include automatic model optimization mechanisms that perform periodic training and fine-tuning because waste composition changes over time. The implementation of the proposed model becomes more effective for real-world applications by combining it with IoT sensors, robotics, and mobile application technology for enhanced waste sorting efficiency. Waste bins that use IoT technology feature cameras and sensors to take pictures, which the proposed system classifies in real time. Smart waste receptacles provide immediate user feedback about correct waste disposal and create recycling process optimization data for waste management organizations. Waste trends can be accessed through cloud-based integration for distant decision-making, as well as remote performance monitoring.

Recycling plants benefit from waste sorting improvement through robotic arms, which operate under AI-based classification models. The model serves as a navigational tool, which enables robotic mechanisms to detect waste items while enabling them to perform the tasks of picking the materials before sorting them into correct receptacles. These systems work together with human operators, as they enhance operational output and decrease the number of mistakes made by human workers during sorting operations. Real-time decisions through robotic systems become possible with Edge AI implementation because cloud-based processing is eliminated, which results in reduced latency.

The mobile application uses the proposed model for users to consult waste item classification outcomes through their smartphone cameras automatically after photographing them. The applications will teach people proper disposal methods while developing responsible waste management behaviours in users. The application features an interface that enables waste collection agencies to create routing plans for waste pickup based on current waste placement information.

Lightweight versions of the model system need to be created to function effectively on embedded devices. A real-time self-learning system should be installed for adaptive waste pattern recognition throughout continuous operations. The system should develop compatibility with existing waste management structures to ensure smooth integration between components. Performing tests with the model in smart cities, alongside industrial waste sorting facilities, will determine its operational effectiveness in actual conditions.

Necessary advancement in waste management technology, along with effective solutions to encountered challenges, will guide the proposed garbage classification system toward efficient, automated, and sustainable waste solutions.

6.6 Conclusion

The experimental results show that this model solves fundamental problems in garbage identification through efficient solutions for handling unbalanced classes, alongside subpar feature acquisition techniques found in ResNet50 and MobileNetV2 models. The model demonstrated significant advancement in all evaluation measurements by implementing attention mechanisms and GAN-based data augmentation, without losing its capability to execute efficiently at the application level. The model's promising results leave room for additional development, which includes acquiring real-time datasets, improving GAN models for quicker data generation, and implementing lightweight attention modules for Raspberry Pi and mobile devices. The research creates initial groundwork for additional studies in the garbage classification field. The research paths for future development consist of utilizing transformer architecture for enhanced feature extraction, alongside GAN extension using Style GAN or diffusion models, which will lead to the deployment of waste sorting systems with IoT devices for intelligent waste management programs. The research proposal targets two key problems of garbage classification systems by establishing novel benchmarks, which can drive sustainable waste management solutions throughout the world.

Future studies should focus on a few crucial areas to guarantee the effective real-world application of the suggested Attention-Based ResNet50 with GANs for garbage categorization. These include making the model more adaptable to changing waste patterns, improving system interoperability, optimizing it for real-time applications, and carrying out practical tests to confirm its efficacy.

It takes effective computing with low hardware resources to deploy deep learning models on embedded and edge devices, like robotic sorters and IoT-enabled trash cans. Efficient architectures investigating lightweight models like MobileNet, ShuffleNet, or EfficientNet, which are optimized for low-power devices; edge AI deployment for implementing the model on microcontrollers, FPGAs, or TPUs to enable high-speed, real-time waste classification without reliance on cloud-based servers; and model compression techniques employing techniques like pruning, quantization, and knowledge distillation to reduce the model's size and inference time while maintaining classification accuracy should be the main areas of future research. Models for classifying garbage must constantly adjust to the shifting composition of waste brought about by new packaging materials, shifting consumer habits, and seasonal fluctuations. To address this, research can focus on incremental learning for developing algorithms that allow the model to learn from new data without forgetting previous knowledge, ensuring continuous improvement, federated learning for enabling distributed training across multiple waste sorting facilities while preserving data privacy, allowing the model to adapt to diverse environments; and active learning pipelines for integrating human feedback into the model's learning process to refine classifications based on real-world errors and edge cases. Previous waste collection methods need to intersect seamlessly with the model for widespread adoption of this technology. Standardization of waste classification represents the main emphasis, while working to establish global waste classification categories that support current municipal recycling guidelines. The AI model gets connected to IoT-based waste monitoring systems through integration for smart waste bins and centralized waste management dashboards that enable real-time decision-making. Additionally, it allows automation in recycling facilities by implementing the model into robotic sorting systems, autonomous waste-sorting vehicles, and conveyor belt cameras. The deployment of this system demands extensive testing within controlled settings along with real-world situations before extensive use. Future assessment efforts will conduct on-site testing in smart cities to implement the model within smart waste collection systems for waste classification of household waste, and commercial and public space waste. The system should become part of industrial pilot programs to assess how the system performs on automated recycling plants regarding the accuracy of classification, processing speed, and operational benefits assessment. A process of gathering real-world waste images, user feedback, and misclassification cases enables the model to receive additional development. The proposed garbage classification system has the ability to define efficient, automated, and sustainable waste management solutions through future development of AI techniques, combined with IoT and robotics. The established recycling system will help worldwide sustainability initiatives by raising recycling percentages and minimizing landfill output.

References

Keval Doshi (2019). Synthetic image augmentation for improved classification using generative adversarial networks. *State Recognition symposium.* https://doi.org/10.32555/2019.dl.008

Ghayoumi, M. (2023). Deep Convolutional generative adversarial networks (DCGANs). *Generative Adversarial Networks in Practice*, 220-257. https://doi.org/10.1201/9781003281344-8

Javid, B.Doshi, & Mashayekhi, H. (2024). A comparative analysis of using generative adversarial networks in Imbalanced text classification. https://doi.org/10.22541/au.172516226.69469834/v1

Kotwal, J., & Koparde, S. (2024). A conditional generative adversarial network and Yolov5 darknet-based skin lesion localization and classification using an independent component analysis model. https://doi.org/10.2139/ssrn.4750698

Luo, L., Cai, J., Chen, Y., Fu, X., & Jiang, H. (2023). The missing value imputation method based on multi-model self-attention generative adversarial networks. https://doi.org/10.2139/ssrn.4555905

Qian, H., Geng, Y., & Wang, H. (2024). Lithology identification based on ramified structure model using generative adversarial network for Imbalanced data. https://doi.org/10.2139/ssrn.4760987

Hualin Chen, Jianan Wei, Haisong Huang, Yage Yuan, Jiaxin Wang, Review of imbalanced fault diagnosis technology based on generative adversarial networks, *Journal of Computational Design and Engineering*, Volume 11, Issue 5, October 2024, Pages 99–124, https://doi.org/10.1093/jcde/qwae075

Sinha, S., Thomopoulos, S. I., Lam, P., Muir, A., & Thompson, P. M. (2021). Alzheimer's disease classification accuracy is improved by MRI harmonization based on attention-guided generative adversarial networks. https://doi.org/10.1101/2021.07.26.453862

Sutedja, I. (2020). Imbalanced data classification using auxiliary classifier generative adversarial networks. *International Journal of Advanced Trends in Computer Science and Engineering*, *9*(2), 1068-1075. https://doi.org/10.30534/ijatcse/2020/26922020

Wang, A., & Zhu, R. (2023). Unpaired bone marrow smears virtual staining using content and attention-guided generative adversarial networks: UBMSVStain using content and attention-guided generative adversarial networks. *Proceedings of the 2023 4th International Symposium on Artificial Intelligence for Medicine Science*, 676-681. https://doi.org/10.1145/3644116.3644228

Arora, M., Ahmad, V., Kumar, R., & Singh, R. (2025). The Impact of Industry 4.0 Technologies on Operational Excellence Methods: Application of Fuzzy Topsis. Engineering Management Journal, 1-12.

Ahmad, V., Arora, M., Sayal, A., & Kumar, R. Exploring Trends in E-Business and Sustainable Finance: An Integrated Analytical Approach. Frontiers in Sustainable Cities, 7, 1549874.

Devi, P., Desai, S. S., Kumar, R., Naruka, M. S., & Tripathi, V. (2024, December). A Strategic Innovations for Stock Market Optimization by Using Data Science & Explainable AI in High Frequency Trading. In 2024 International Conference on IoT Based Control Networks and Intelligent Systems (ICICNIS) (pp. 1510-1515). IEEE.

Kumar, R., Goel, R., Singh, T., Mohanty, S. M., Gupta, D., Alkhayyat, A., & Khanna, R. (2024). Sustainable Finance Factors in Indian Economy: Analysis on Policy of Climate Change and Energy Sector. Fluctuation & Noise Letters, 23(2).

Kumar, R., Kandpal, B., & Ahmad, V. (2023). Industrial IoT (IIOT): Security threats and countermeasures. *2023 International Conference on Innovative Data Communication Technologies and Application (ICIDCA)*, 829–833. IEEE

https://www.kaggle.com/datasets/mostafaabla/garbage-classification

Song, Y., He, X., Tang, X. *et al.* DEEPBIN: Deep Learning Based Garbage Classification for Households Using Sustainable Natural Technologies. *J Grid Computing* **22**, 2 (2024). https://doi.org/10.1007/s10723-023-09722-6

7

AI-powered Waste Management: Advancing Sustainability and a Circular Economy Through Intelligent Systems

Swarna Prabha Jena[1] and Bijay Kumar Paikaray[2]

[1]Department of Electronics and Communication Engineering, Centurion University of Technology and Management, India
[2]Department of Computer Science and Engineering, Siksha 'O' Anusandhan (Deemed to be University), India
E-mail: prabha.jena@gmail.com; bijaypaikaray87@gmail.com

Abstract

Waste management has become a worldwide problem that requires creative ideas to maintain sustainability and create a circular economy (All noman et al., 2022). Conventional models are failing more and more to address the growing waste levels and their effects on ecology. In this era, traditional models are to be converted into modern technologies for effectively tackling waste. Here, the focus will be on automated sorting, recycling enhancement, predictive analytics, and ethical issues. The use of artificial intelligence (AI) in waste management is also highlighted. It investigates practical applications, including robotics combined with machine learning (ML), and AI-powered optical sorting devices for effective e-waste management. This also focuses on case studies based on the Internet of Things (IoT) and machine technologies in optimizing operations, lowering costs, and minimizing environmental risk. The chapter also closely reviews ethical issues and policy frameworks, influencing waste management AI acceptance. This study offers researchers, policymakers, and companies an opportunity to work together to bridge the knowledge gap, so they may cooperatively improve sustainable

DOI: 10.1201/9788743806905-7

waste management techniques. These developments offer a radical change toward a future based on ecologically robust resources.

Keywords: Artificial intelligence, waste management, sustainability, technology, IoT.

7.1 Introduction

Managing waste is a worldwide problem that needs new technological answers to make it efficient and support a circular economy for the effective future, as shown in Figure 7.1. Much technology has come up, starting from AI to IoT, which solves these problems, make waste management better, and help reach global sustainability goals. This section deals with the requirements, the role of cutting-edge technology contributions, and the circular economy (All noman et al., 2022); use cases, conflicts, and ethical issues are discussed in the subsequent sections.

7.1.1 Requirement for effective waste management

AI-enabled waste management offers a transformative opportunity to address one of the global challenges (Kaza et al., 2018). The integration of AI, IoT, and ML into waste management strategies has the potential to enhance efficiency, sustainability, and cost-effectiveness to extraordinary levels (Dutta & Goel, 2021). Achieving the above vision needs collaborative efforts among policymakers, researchers, and industries, as seen by the real steps taken to accomplish it. By reducing environmental effects and overcoming resource use, AI-driven waste management systems can greatly help toward a circular economy. With international collaboration and ongoing technological development promoting a more sustainable future (Esmaeilian et al., 2018) AI will be key in this change.

7.1.2 The role of AI in waste management

Artificial intelligence (AI) offers innovative solutions to modern waste management challenges by leveraging automation, predictive analytics, and data-driven decision-making (Lanzalonga et al., 2024). The AI-based solution [12] is capable of handling large data intensities and facilitating accurate classification of different categories of wastes. Other predictive models utilize machine learning to predict patterns of waste generation, hence also giving

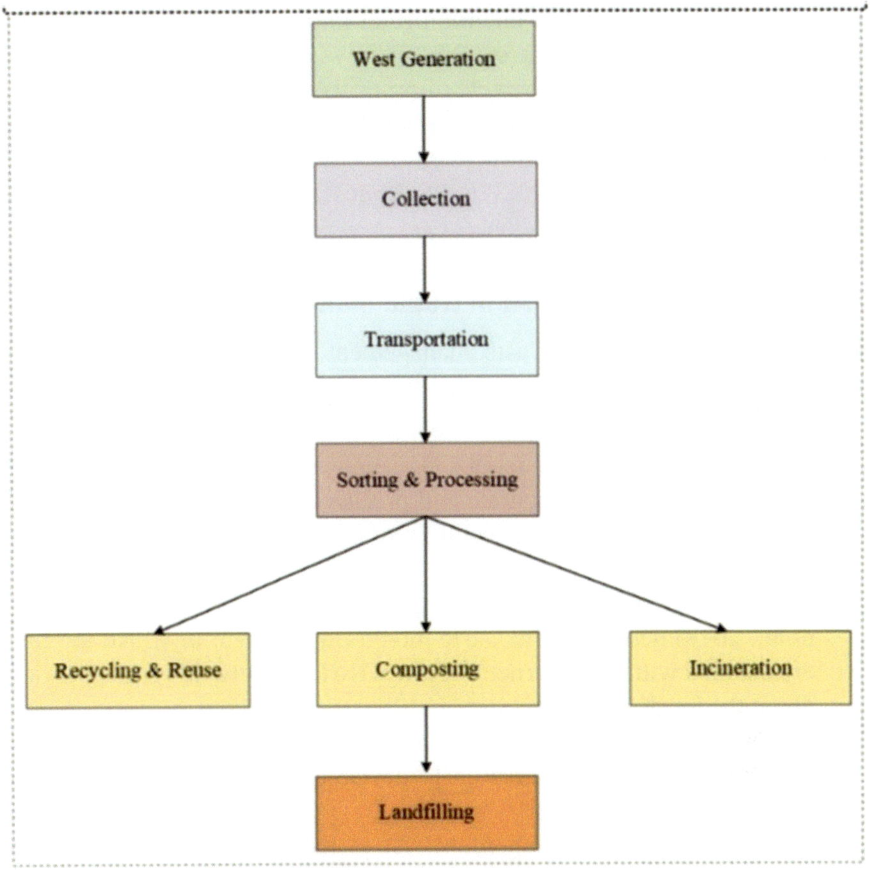

Figure 7.1 Traditional waste management scenario.

recommendations on strategies for waste reduction, which directly trace back to SDG 12. Efficient in the sense of remedying the inefficiency of classical systems, AI facilitates intelligent waste collection, recycling, and monitoring, and sets out to accomplish sustainability goals [10].

7.1.3 AI and the birth of the circular economy

In the definition of a circular economy championed by Ellen (2019), attention is placed on designing out waste from the system through reuse, recycling, and remanufacturing of materials. AI, by detecting and precisely naming materials that are recyclable, abides by the process of resource efficiency, at the same time reducing its environmental impact. AI can be used in any municipal waste management system. This will make it possible for the organization to extend the life cycle of products, remove waste, and comply with the conditions for achieving SDG No. 9 and No. 13, which are founded on industry, innovation, infrastructure, and climate action respectively. As highlighted by Chidepatil et al. (2020), AI accelerates the transition toward sustainable resource consumption.

7.1.4 Major use cases of AI in waste management

The applications of AI in waste management are numerous and cover a variety of key sectors and domains, including:

- **Automated waste sorting**: The automated waste sorting system is designed (Gundupalli et al., 2017). Computer vision and ML geometry could be used for the accurate and precise identification of objects and this would increase the recycling rate because more materials can be reused.
- **Smart waste monitoring**: The authors (Hannan et al., 2015, Hannan et al., 2018) described the waste monitoring system, in which the bins are enabled with the Internet of Things (IoT) and provide real-time data about the levels of waste.
- **Predictive analytics**: There are ML models used to examine and analyze historical waste formation to devise efficient solutions for waste reduction.
- **Hazardous waste management**: AI in the identification and secure separation of hazardous waste, guaranteeing that environmental requirements are adhered. It is well-acknowledged that these applications are of utmost significance in terms of achieving operational efficiency and promoting behaviors that are environmentally appropriate in terms of waste management.

7.1.5 Conflicts and ethical issues

The implementation of artificial intelligence in waste management is not without its difficulties. Several issues need to be addressed, including high implementation costs, concerns around data privacy, and technological hurdles. In addition, (Dutta & Goel, 2021) emphasize the importance of ensuring the ethical deployment of AI to prevent biases in decision-making and to guarantee transparency (Gaur & Gupta, 2020). To overcome these obstacles and develop the proper application of artificial intelligence for waste management solutions, collaborative initiatives including governments, companies, and researchers are completely necessary.

7.1.6 Chapter's scope

This chapter explores the case studies of real-world practice, technological diffusion, and policies for AI in sustainability and the a circular economy. The authors (Esmaeilian et al., 2018) discuss why, where, when, and how the policies are applied in the area of waste management? This chapter will enable you to see precisely how policies and technological advancements in AI help sustainability and the circular economy maximize the potential for making economics sustainable. Islam & Huda (2019) attempt to fill the information gaps with interpretations of the ideas presented, which should guide researchers, policymakers, and organizations to new integrative directions and approaches toward efficient waste management.

7.2 AI in Waste Sorting and Recycling

7.2.1 Automated waste sorting

Esmaeilian et al. (2018) state that AI-based waste-sorting systems fundamentally renovate traditional waste-sorting methods by using computer vision and ML to detect waste materials with high accuracy. Such systems operate based on large data volumes and training fairly, so there remain concerns about data privacy. For example, the datasets that are used to train the model have hypothetically sensitive geolocated images of waste collection places. In order to handle this problem and ensure that they are in compliance with data protection rules such as the General Data Protection Regulation (GDPR), organizations make use of secrecy and secure data storage. The applications, such as the AI-powered waste-sorting systems developed by ZenRobotics, demonstrate that there is no mutual exclusion between secure data management strategies and technological advancements.

7.2.2 Robotics in waste management

There are robotic systems developed that are equipped with AI-based image recognition technology that have improved the efficiency of waste sorting, as stated by Hossain & Roy (2020). Despite this, algorithmic discrimination continues to be a big concern. If the majority of the training samples consist of garbage from metropolitan areas, for instance, the AI would not work well in rural locations and might even lead to inefficiencies. To mitigate this issue, researchers are focusing their efforts on broadening training datasets to integrate garbage samples from a variety of socioeconomic and geographical locales. One notable system developed by AMP Robotics taken as an example is the robotic arm-based sorting system. This system maintains efficiency across regions by adapting its algorithms to different wastes.

7.2.3 Smart recycling bins

Intelligent recycling bins employ technology such as sensors, artificial intelligence, and the Internet of Things to optimize collection routes and save operating expenses by delivering real-time data on diverse garbage types and quantities (Hannan et al., 2015, Hannan et al., 2018). These bins, however, mitigate prevalent issues, including contamination, poor sorting, and privacy concerns stemming from the regular collecting of user behavior data by these systems. Numerous real-time examples include Bin-e, Racoon, iTrash, and Convwaste. To address this issue, firms such as Bigbelly Solar have implemented robust data governance protocols. These standards ensure the security of user data, which is largely employed to improve the effectiveness of waste management. These projects promote and exemplify the worldwide dedication to bolstering public trust and to facilitating the broader use of modern recycling systems.

7.2.4 AI-powered optical sorting systems

AI-powered optical sorting automates garbage and recycling. This system often uses high-speed cameras, AI algorithms, and smart sensors with mechanical sorters. The authors (Hossain & Roy, 2020) state that the optical sorting systems use AI and hyperspectral imaging to identify, sort, segregate, and assess waste composition in real time. This technology improves sorting accuracy and efficiency and promotes a circular economy initiatives. Even waste's dangerous materials, pathogens, and chemicals damage humans. Thus, intelligent technology solves the problem and enhances the workplace. Technology improves material recovery, efficiency, operational costs, and resource reuse. AMP Robotics, Zen Robotics, and Tomra Solutions efficiently sort. AI decisions made in secret can damage public trust. To overcome this, companies utilize explainable AI (XAI) frameworks to understand sorting decisions. TOMRA Sorting Solutions' optical sorting systems use XAI for accountability and stakeholder confidence.

7.2.5 Deep learning in waste classification

DL has developed effective waste classification capabilities. By analyzing massive datasets of garbage photos, DL models like CNNs help classify waste (Jain & Singhal, 2020). These systems classify trash without human intervention. Biases in these datasets can skew results, and dataset availability and computational resources can impair recycling efficiency. Researchers are using adversarial training to find and fix AI model biases. A World Economic Forum case study shows how such approaches have enhanced AI-driven garbage classification systems, assuring equitable and effective recycling. These issues can be addressed by federated learning and edge computing.

7.2.6 AI in E-waste management

According to Islam & Huda (2019) the management of electronic trash (e-waste) presents distinct obstacles owing to its toxic constituents and intricate recycling necessities. AI-driven systems, exemplified by Apple's Daisy robot, have shown the capability to automate the disassembly of e-waste and the recovery of materials. Nonetheless, ethical issues emerge about the environmental ramifications of AI hardware. To mitigate this issue, corporations are investigating sustainable AI techniques, like the implementation of energy-efficient algorithms (Kumar & Samadder, 2017) and the recycling of AI hardware components, to reduce their ecological impact.

7.2.7 Challenges in AI-powered waste sorting

Modern waste management systems have increased efficiency. AI-powered waste sorting faces challenges like variability in waste materials based on shape and size, high sensor and hardware investment, integration complexities, energy consumption issues (Kumar & Samadder, 2017), and ethical concerns despite its transformative potential [6] Data privacy concerns can hinder AI adoption. Organizations are using hybrid AI robots and blockchain technologies to secure and transparently share data. Equitable AI implementation requires continual model training and validation to meet emerging capabilities. Collaboration among government, industry, and academic is needed to solve these problems.

7.2.8 Future prospects of AI in waste sorting and recycling

Ellen (2019) reports that automation, robotics, and sensors are advancing waste sorting and recycling technology. The future (Ghisellini et al., 2016) depends on ethical AI development. Data privacy and algorithmic fairness are prioritized for a more sustainable, economical, and environmentally friendly system. AI-driven drones for garbage surveillance and blockchain technology demonstrate the promise of innovative recycling solutions to boost efficiency and sustainability. These strategies reduce environmental degradation by encouraging stakeholder participation. Waste management may completely use AI to create a circular economy.

The optimal description of this value depends on the system's characteristics and the type of equipment. As an example, for modulation and coding techniques in wireless communications, spectral efficiency is a common measure. For electronic components, the ratio of joule per bit best describes performance. In telecommunication networks and data centers, the ratio of watts consumed over the Gbps of data processed is preferred. An absolute energy (Kumar & Samadder, 2017) efficiency metric is introduced, named as dBε.

7.3 AI in Waste Monitoring and Management

7.3.1 AI-based smart waste monitoring systems

AI-driven smart waste monitoring systems use IoT sensors, machine learning algorithms, and cloud-based analytics to revolutionize waste management. These systems monitor waste levels, composition, and disposal in real-time. Authors (Hannan et al., 2015, Hannan et al., 2018) found that AI-enabled smart bins with fill-level sensors can determine garbage kind and quantity. These containers inform users when they are full, reducing overflow and improving operating efficiency. These devices decrease wasteful pickups, saving fuel and carbon emissions for municipalities and garbage management corporations. AI-based monitoring reveals trash generation hotspots for targeted interventions and better resource allocation.

7.3.2 Predictive analytics for waste management

Modern garbage management relies on AI-powered predictive analytics. AI systems can accurately predict garbage production trends by examining past data and patterns. These projections allow policymakers and waste management authorities to optimize garbage collection schedules, identify recycling facility locations, and conduct targeted awareness programs (Gupta & Jain, 2020). Urban AI technologies can estimate seasonal garbage creation, helping cities spend resources during peak periods. This data-driven strategy improves operational efficiency and waste reduction, supporting sustainability goals.

7.3.3 AI-enhanced waste tracking and logistics

AI has made transportation more efficient, which is essential to waste management. AI-powered waste tracking systems use IoT and blockchain to track waste from generation to disposal. Hannan et al., 2015 and Hannan et al., 2018 report that these systems can track waste, check disposal restrictions, and detect illegal dumping. AI analyzes real-time traffic, weather, and bin-fill data to improve waste collection routes. Fuel usage, delays, and operational costs decrease. AI-driven route optimization has reduced automobile emissions and improved rubbish collection in cities.

7.3.4 Automated hazardous waste detection

Because improper disposal can harm the environment and health, hazardous waste management is difficult. AI-powered technologies are being used to detect and classify hazardous waste. Hossain & Roy (2020) report that these systems can precisely identify heavy metals, dangerous chemicals, and radioactive materials using advanced deep-learning models and chemical analysis techniques. Segregation is automated using AI, improving hazardous waste management and regulatory compliance. Some AI-based systems use spectroscopic methods and machine learning to analyze waste contents in real time, enabling safe and quick management of hazardous wastes.

7.3.5 Challenges and future prospects in AI-based waste monitoring

Gupta & Jain (2020), state that AI-based waste monitoring confronts various hurdles despite its transformational promise. High implementation costs and solid technology infrastructure might be hurdles in resource-constrained countries. Data privacy and AI system security remain major problems. Technology like AI, IoT, and cloud computing should lower costs and increase accessibility. The future [8] of AI in garbage monitoring involves AI-powered drones for aerial waste assessment, improved image recognition models for real-time sorting, and edge computing for quicker data processing. AI-driven garbage monitoring and management can help achieve global sustainability targets by overcoming present limits.

7.4 AI Integration with IoT and MI

7.4.1 The role of IoT in waste management

Waste management relies on the Internet of Things (IoT) to connect devices and systems. Gaur and Gupta (2020) reported that smart bins and garbage pickup vehicles use IoT-enabled sensors to monitor waste levels, composition, and disposal routines in real-time. A central system processes this data to make educated judgments using AI. IoT sensors can alert collection personnel when a bin is becoming full, facilitating prompt garbage removal, and preventing overflow. AI-integrated IoT system can improve garbage collection routes, saving fuel, operational expenses, and carbon footprint (Figure 7.3).

7.4.2 AI and machine learning in predictive waste management

Jain & Singhal (2020) examine how ML, which is a subset of AI, analyzes historical data to forecast future waste generation trends in predictive waste management. AI systems recognize trash production and disposal patterns, helping enterprises and municipalities prepare ahead. Predicting a rise in quantity of waste during vacations or festivals allows optimal resource use. These predicted insights help reduce landfill dependence, increase recycling, and adopt sustainable waste reduction methods.

7.4.3 Real-time waste monitoring and decision-making

Real-time decision-making and monitoring of the waste is enabled through IoT, AI, and data analytics. Mittal et al. (2021) use AI-powered IoT systems to monitor garbage across collection stations, landfills, and recycling centers. Smart sensors monitor the waste, its type, gases, and the environment which affects the normal human. The sensor's data are remotely collected and displayed on mobile and web apps. This data on the webpage gives the information and provides alerts and instant decisions will be taken. Waste management authorities may make swift, data-driven (Lanzalonga et al., 2024) choices using this real-time data. If a landfill's methane levels rise, the system may notify authorities to mitigate concerns. This also helps prevent pollution, increase efficiency, and reduce operational costs. The real-world systems which are already developed are Bigbelly smart waste bins, Enevo smart sensors, and Bin-e smart waste system. These systems not only enhance tracking accuracy but also enable autonomous waste collection making the system more transparent.

7.4.4 AI-enabled waste-sorting automation

AI and IoT improve garbage sorting through automated categorization. AI algorithms classify garbage based on IoT sensor data such as material kind and condition. Automation eliminates pollution in recycling streams and boosts recovery. It is found that AI-enabled robotic arms connected to IoT can precisely recognize and sort plastics, metals, and paper (Hassan et al., 2022 & Yang et al., 2020). These innovations maximize resource use and reduce waste, promoting a circular economy concepts.

7.4.5 Challenges and future prospects

Integration of AI, IoT, and machine learning in waste management is difficult due to high implementation costs, data security issues, and system compatibility. These difficulties demand creative solutions like edge computing for speedier data processing and blockchain for transparent data management. In [7], real-time monitoring systems, autonomous garbage pickup trucks, and AI-driven resource optimization analytics. Governments, industry, and technology providers may collaborate to fully achieve AI-IoT integration's potential, contributing to sustainable waste management and global environmental goals [21].

7.5 Policy and Ethical Considerations in AI Waste Management

7.5.1 Regulatory frameworks for AI in waste management

Governments worldwide are realizing the necessity to regulate AI-driven waste management systems for safety and environmental compliance. AI systems that handle sensitive waste management data must comply with the EU's General Data Protection Regulation (GDPR) for data protection and security. The circular economy policy frameworks stress openness and sustainability by integrating AI. The Indian Swachh Bharat Mission uses AI to improve trash management. Unfortunately, these guidelines seldom provide explicit AI deployment instructions, which might confuse when implemented. Governments might fix this by establishing consistent AI integration norms that emphasize data control, EIAs, and ethical AI.

7.5.2 Ethical challenges in AI waste management

Data privacy, algorithmic prejudice, and worker displacement are ethical problems with AI-driven trash management. AI algorithms that train on massive datasets may reveal sensitive data like geotagged rubbish pickup places. To avoid this, organizations must use strong data anonymization and comply with GDPR; another major concern is algorithmic prejudice. AI models trained on undiversified datasets may underperform in specific locations or populations, resulting in unequal waste management. Continuous model validation and various datasets are needed to assure fairness and accuracy.

7.5.3 Balancing AI efficiency with human labor

AI, blockchain, and ML IoT have been discussed extensively since the preceding section, raising worries regarding waste sorting and recycling automation job displacement in the waste management business. As AI improves efficiency, it should supplement human labor. Also, the human touch is invaluable. Human workers may focus on supervision and decision-making, while AI handles repetitive and dangerous duties. However, people need skills and training to eliminate it. Yes, the AI system can provide biased results, hence humans are essential for fairness and transparency. To prepare people for positions in AI-integrated waste management system, governments and companies must engage in workforce reskilling or upskilling.

7.5.4 Gaps in policy and future directions

Oh et al. (2018) highlight that despite existing regulations, significant gaps remain in policy frameworks for AI adoption in waste management. For example:

- **Lack of standardization**: There is no universal standard for AI deployment in waste management, leading to inconsistencies in implementation across regions.
- **Limited focus on ethical AI**: Current policies often overlook the ethical implications of AI, such as transparency in decision-making and accountability for errors.
- **Insufficient incentives for adoption**: High implementation costs deter organizations from adopting AI technologies. Policies offering financial incentives, such as tax breaks or subsidies, could encourage wider adoption.

It is necessary for the policies and structures of the next generation to highlight:

- **Sustainability goals**: Facilitating the alignment of artificial intelligence-driven waste management programs with global sustainability goals, such as the sustainable development goals (SDGs) established by the United Nations.
- **International collaboration:** Establishing cross-border partnerships to share best practices and develop standardized AI protocols.
- **Public-private partnerships**: It is important to foster collaboration between research institutions, industry, and governments to promote innovation and guarantee that everyone has equal access to artificial intelligence technologies.

7.6 Future Prospects and Conclusion

7.6.1 Actionable recommendations for policymakers

Policymakers play a pivotal role in driving the adoption of AI-powered waste management systems. To align with global (Triassi et al., 2015) efforts like the UN SDGs, the following steps can be undertaken:

- Develop standardized frameworks: To ensure that artificial intelligence is ethically used in waste management, it is necessary to establish clear and uniform criteria that cover data protection, transparency, and sustainability.
- Provide financial incentives: Offer subsidies, tax incentives, or grants to encourage municipalities and organizations to invest in AI technologies.
- Promote public-private partnerships: Facilitate collaboration between governments, industries, and research institutions to develop and implement AI-powered solutions.
- Focus on capacity building: Invest in training programs to upskill the workforce, ensuring they can operate and maintain AI-enabled waste management systems.

Policymakers can contribute to SDG 11 (Sustainable cities and communities) and SDG 12 (Responsible consumption and production) by creating an environment that is favorable to the widespread adoption of AI through the implementation of these measures, as elucidated (Esmaeilian et al., 2018).

7.6.2 Recommendations for researchers

In the field of waste management based on artificial intelligence, researchers have the opportunity to pioneer breakthroughs and address concerns that already exist. Among the most significant concerns are:

- **Advancing machine learning models**: Develop more accurate and unbiased AI models for waste classification and sorting, tailored to diverse waste streams and geographies.
- **Enhancing AI-IoT integration**: Focus on seamless integration of IoT devices with AI systems for real-time data collection and decision-making.
- **Exploring a circular economy applications**: Investigate how AI can optimize resource recovery and recycling within the circular economy framework (Cucchiella et al., 2017).
- **Studying ethical implications**: Conduct in-depth research on the ethical dimensions of AI, such as algorithmic bias, transparency, and societal impacts.

Through these efforts, researchers can support SDG 9 (Industry, innovation, and infrastructure) and SDG 13 (Climate action) by advancing sustainable technologies for waste management.

7.6.3 Recommendations for industries

It is essential that industries, which are major players in waste management, take proactive steps to integrate AI technology, as indicated in (Gupta & Jain, 2020) research. These include a few of them:

- **Investing in innovation**: Allocate resources for the development and deployment of AI-powered waste management systems, focusing on automation and efficiency.
- **Adopting sustainable practices**: Leverage AI to achieve sustainability goals, such as minimizing waste, improving recycling rates, and reducing operational costs.
- **Collaborating with stakeholders**: Partner with governments, researchers, and other industries to share best practices and address common challenges in AI deployment.
- **Implementing transparent processes**: Use blockchain technology alongside AI to ensure traceability and accountability in waste disposal and recycling operations.

By embracing these strategies, industries can play a significant role in meeting SDG 8 (Decent work and economic growth) and SDG 17 (Partnerships for the goals).

7.6.4 Looking ahead: Future innovations and global impact

The future of AI-powered waste management holds immense promise. Innovations are done such as AI-driven drones for aerial waste monitoring, enhanced robotics for precise sorting, and blockchain-enabled systems. These innovations are used for transparent waste tracking on the horizon. These AI and IoT technologies will not only improve operational efficiency but also contribute to environmental conservation and resource optimization. The collaborative hard work will be essential to overcome and address the challenges such as high implementation costs and data privacy apprehensions. Some real work examples are already discussed in the above sections which make this system more environmentally friendly. By nurturing partnerships across borders, stakeholders can share knowledge, pool resources, and promote the adoption of AI technologies globally (Kaza et al., 2018). Aligning these initiatives with the UN SDGs ensures that AI-powered waste management contributes meaningfully to a cleaner, more sustainable world.

7.7 Conclusion

Waste management is driven by AI and presents a novel approach to addressing a globally significant problem. When used in waste management systems, AI, IoT, and ML have the potential to significantly improve efficiency, sustainability, and cost-effectiveness (Consonni & Viganò, 2012). To accomplish this objective, policymakers, researchers, and businesses must work together. This is something that is proved by the current strategies that have been discovered to enable its execution. Waste management systems that are powered by AI have the potential to significantly ease the transition to a circular economy by letting down the impact on the environment and improving resource efficiency. To achieve a more sustainable future, it will be important to continue the development of technology in conjunction with international cooperation. Artificial intelligence will play a significant part in this transition.

References

All noman, A., Akter, U. H., Pranto, T. H., & Haque, A. B. (2022). Machine learning and artificial intelligence in a circular economy: A bibliometric analysis and systematic literature review. arXiv preprint arXiv:2205.01042.

Bianchini, A., Rossi, J., & Pellegrini, M. (2020). Artificial intelligence for sustainable waste management: A critical review. Journal of Environmental Management, 271, 111-123.

Chidepatil, A., Bindra, S., Kulkarni, M., Qazi, M., & Khetrapal, S. (2020). From trash to cash: How blockchain and multi-sensor-driven artificial intelligence can transform a circular economy of plastic waste. Administrative Sciences, 10(2), 23.

Consonni, S., & Viganò, F. (2012). Waste gasification vs. conventional waste-to-energy: A comparative evaluation of two commercial technologies. Waste Management, 32(4), 653-666.

Cucchiella, F., D'Adamo, I., & Gastaldi, M. (2017). Sustainable waste management: Waste to energy plant as an alternative to landfill. Energy Conversion and Management, 131, 18-31.

Dutta, P., & Goel, S. (2021). Artificial intelligence in solid waste management: A survey of the state of the art. Waste Management & Research, 39(6), 743-757.

Ellen MacArthur Foundation. (2019). Artificial intelligence and the circular economy: AI as a tool to accelerate the transition. Ellen MacArthur Foundation.

Esmaeilian, B., Wang, B., Lewis, K., Duarte, F., Ratti, C., & Behdad, S. (2018). The future of waste management in smart and sustainable cities: A review and concept paper. Waste Management, 81, 177-195.

Gaur, N., & Gupta, R. (2020). Artificial intelligence in waste management to achieve sustainable development goals: A study of India. Journal of Cleaner Production, 276, 124-131.

Ghisellini, P., Cialani, C., & Ulgiati, S. (2016). A review on a circular economy: The expected transition to a balanced interplay of environmental and economic systems. Journal of Cleaner Production, 114, 11-32.

Gundupalli, S. P., Hait, S., & Thakur, A. (2017). A review on automated sorting of source-separated municipal solid waste for recycling. Waste Management, 60, 56-74.

Gupta, S., & Jain, A. (2020). Artificial intelligence applications in solid waste management: A review. Waste Management & Research, 38(5), 487-508.

Hannan, M. A., Hasan, M. N., Begum, R. A., Basri, H., Hussain, A., & Scavino, E. (2015). Review of baseline studies on energy policies and indicators in Malaysia for future sustainable energy development. Renewable and Sustainable Energy Reviews, 49, 1125-1137.

Hannan, M. A., Hassan, M. K., Begum, R. A., Basri, H., & Hussain, A. (2015). Optimization of solid waste collection and transportation routes incorporating GIS-based systems: Case study of Dhaka city. Environmental Monitoring and Assessment, 187(4), 199.

Hannan, M. A., Hossain Lipu, M. S., Hussain, A., & Mohamed, A. (2018). A review of internet of energy-based building energy management systems: Issues and recommendations. IEEE Access, 6, 38997-39014.

Hannan, M. A., Mujibu, M. A., Begum, R. A., & Basri, H. (2011). A review on technologies and their usage in solid waste monitoring and management systems: Issues and challenges. Waste Management, 31(8), 1889-1899.

Hannan, M. A., Zailani, S., Kazi, Z. H., & Uddin, M. N. (2015). Intelligent waste management system with RFID technology. International Journal of RFID Security and Cryptography (IJRFIDSC, 4(1), 1-10.

Hassan, M., Ali, F., & Zhang, Y. (2022). AI-Driven Waste Sorting System for Smart Cities: A Deep Learning Perspective. Waste Management & Research, 40(3), 233-245.

Hossain, M. S., & Roy, P. (2020). Application of artificial intelligence in waste management: A review. Journal of Environmental Management, 271, 111-123.

Islam, M. T., & Huda, N. (2019). Assessing the recycling potential of "unregulated" e-waste in Australia. Resources, Conservation and Recycling, 145, 277-286.

Jain, S., & Singhal, S. (2020). Application of artificial intelligence in sustainable waste management practices: A review. Journal of Cleaner Production, 277, 124-130.

Kaza, S., Yao, L., Bhada-Tata, P., & Van Woerden, F. (2018). What a waste 2.0: A global snapshot of solid waste management to 2050. World Bank Publications.

Kumar, A., & Samadder, S. R. (2017). A review on technological options of waste to energy for effective management of municipal solid waste. Renewable and Sustainable Energy Reviews, 78, 888-903.

Lanzalonga, F., Marseglia, R., Irace, A., & Biancone, P. P. (2024). The application of artificial intelligence in waste management: Understanding the potential of data-driven approaches for the circular economy paradigm.

Management Decision. Advance online publication. https://doi.org/10.1108/MD-10-2023-1733

Mittal, G., Singh, A., & Sharma, R. (2021). Deep Learning for Waste Classification: A CNN-Based Approach. Journal of Environmental AI, 15(2), 45-56.

Arora, M., Ahmad, V., Kumar, R., & Singh, R. (2025). The Impact of Industry 4.0 Technologies on Operational Excellence Methods: Application of Fuzzy Topsis. Engineering Management Journal, 1-12.

Ahmad, V., Arora, M., Sayal, A., & Kumar, R. Exploring Trends in E-Business and Sustainable Finance: An Integrated Analytical Approach. Frontiers in Sustainable Cities, 7, 1549874.

Devi, P., Desai, S. S., Kumar, R., Naruka, M. S., & Tripathi, V. (2024, December). A Strategic Innovations for Stock Market Optimization by Using Data Science & Explainable AI in High Frequency Trading. In 2024 International Conference on IoT Based Control Networks and Intelligent Systems (ICICNIS) (pp. 1510-1515). IEEE.

Kumar, R., Goel, R., Singh, T., Mohanty, S. M., Gupta, D., Alkhayyat, A., & Khanna, R. (2024). Sustainable Finance Factors in Indian Economy: Analysis on Policy of Climate Change and Energy Sector. Fluctuation & Noise Letters, 23(2).

Kumar, R., Kandpal, B., & Ahmad, V. (2023). Industrial IoT (IIOT): Security threats and countermeasures. 2023 International Conference on Innovative Data Communication Technologies and Application (ICIDCA), 829–833. IEEE

Oh, T. H., Hasanuzzaman, M., & Rahim, N. A. (2018). Energy policy and sustainable development in Malaysia: Issues and challenges for the future. Renewable and Sustainable Energy Reviews, 81, 3021-3031.

Triassi, M., Alfano, R., Illario, M., Nardone, A., Caporale, O., & Montuori, P. (2015). Environmental pollution from illegal waste disposal and health effects: A review on the "triangle of death." International Journal of Environmental Research and Public Health, 12(2), 1216-1236.

Yang, J., Wang, T., & Liu, H. (2020). Transfer Learning in Deep Learning-Based Waste Classification: A Review and Case Study. Journal of Sustainable AI, 8(1), 78-92.

8

Intelligent Systems in Waste Management: A Machine Learning Approach

Devra S. J.[1], Prity Kumari[2], Kishan M. Kaneriya[3], and Raj S. R.[4]

[1]Department of Agricultural Statistics, College of Agriculture, J.A.U., India
[2]College of Horticulture, A.A.U., India
[3]Vanbandhu Polytechnic in Agriculture, S.D.A.U., India
[4]Department of Agricultural Statistics, College of Agriculture, A.A.U., India
E-mail: snehdevra21@gmail.com

Abstract

This chapter examines the revolutionary ability of machine learning (ML) to revolutionize waste management systems. Classical waste management methods, which are often characterized by inefficiencies, labor, and high operational expenditure, are becoming more and more supplemented or even replaced by data-driven, smart solutions. The chapter discusses how various machine learning methods, including supervised learning, unsupervised learning, and reinforcement learning, are applied to automate garbage sorting, forecast waste generation, optimize collection routes, and enhance recycling efficiency. Real-world case studies illustrate the application of ML in enhancing operating efficiency, reducing costs, and promoting sustainability in urban garbage management systems. However evident the advantages may be, drawbacks such as quality of data, high initial outlay, and change resistance are debated. To conclude, the chapter peeks into the future, where smart cities and a circular economy models interact with machine learning, and even the ethical concern of applying the intelligent systems comes into focus. This chapter discusses a comprehensive overview of the current state and future potential of machine learning applied to waste management, highlighting

DOI: 10.1201/9788743806905-8

the promise of improving both sustainability and efficiency in the waste management sector.

Keywords: Machine learning (ML), waste management, supervised learning, unsupervised learning, reinforcement learning.

8.1 Introduction

8.1.1 Overview of waste management challenges

The environmental issue that is probably the most severe facing the world today is waste disposal. As urban populations increase, the amount of trash that is produced increases linearly which is putting strain on existing management infrastructure. The increasing generation of waste causes environmental and logistical problems for cities and municipalities worldwide. Poor or ineffective waste management practices can lead to pollution and unsightly landfills, the wasting of resources that could be recycled or reused (Arora Madhu et al., 2025). In an era when sustainable development is a global focus area, efficient waste management will be key to achieving sustainability and the long-term health of the environment (Vasim, A et al., 2025).

Separating waste into certain categories such as recyclable, biodegradable, and hazardous waste is crucial to effective waste management. Therefore, when the materials get recycled, contamination can occur and make the process more difficult and costly. Without sorting, useful products can still end up in the landfill, while dangerous materials can still end up damaging the health of human beings and the environment. Traditional approaches for waste sorting rely on manual processes, which are labor-intensive, error-prone, and thus inefficient in recycling and waste treatment systems.

Conventional waste management systems such as collecting, sorting, and disposing are ineffective. These actions also lead to unaffordable operating expenses and more human errors. Moreover, waste management systems are often reactive and respond to trash build-ups rather than countering it or optimizing waste collection and dumping contractions. Such inefficiencies not only affect the environment, but also the economic viability of waste systems in urban and rural areas (Kumar R. et al., 2025).

With growing awareness as well as concern toward environmental pollution along with the adverse effects of needless waste, more emphasis is settled on meeting sustainability targets including reducing waste, increasing recycling, plus creating the round economy. Governments, cities, and companies are trying to create systems that produce little garbage, recycle as much material as possible, and reduce the external harm from trash disposal. Yet reaching these objectives necessitates overcoming major obstacles from logistics issues and enormous infrastructure costs to resistance from all sections of society.

8.2 Introduction to Machine Learning in Waste Management

ML and AI applications are being ever more widely adopted in the industry to improve efficiency, reduce costs, and enable better decision-making. Waste management, in particular, stands to reap huge returns from big data-based models driven by machine learning. Traditional trash management systems, which frequently rely on manual labor and human involvement, can be improved using automation and intelligent algorithms. These advances have the potential to automate essential activities such as garbage classification, predictive analysis for waste creation, and route optimization, hence increasing sustainability and lowering operational costs.

8.2.1 Machine learning basics and relevance to waste management

8.2.1.1 What is machine learning?

Machine learning is a class of techniques that enable systems to learn from data and detect patterns without requiring explicit programming. These models evolve over time as they process enormous datasets and react to new information. Machine learning is defined by its ability to create predictions, automate processes, and even increase forecast accuracy without ongoing human interaction (Gao et al., 2024). Machine learning can be categorized into three primary types, each relevant to waste management:

- **Supervised learning**: This method involves training a model on a labeled dataset, with the input data linked with the proper output. The model learns to predict the output given the input. In trash management, supervised learning is used to classify waste categories (such as plastics, paper, and metal) using labeled photographs or sensor data.
- **Unsupervised learning**: Unsupervised learning, unlike supervised learning, operates on unlabeled data. The algorithm identifies patterns

or clusters in the data. Unsupervised learning may assist in trash management by clustering waste streams, identifying similar waste types, and improving recycling operations.

- **Reinforcement learning**: Reinforcement learning has been categorized as a sort of machine learning where an agent learns how to make decisions by interacting with its environment and receiving feedback on its actions. Reinforcement learning for waste management can be used for optimizing waste sortation systems or waste collection routes by receiving real-time data and learning over the time to achieve peak efficiency (Kumar et al., 2023)

8.3 Why ML in Waste Management?

- Automation: Waste management processes, from sorting to classification, have been historically resourceful and susceptible to errors. Machine learning takes these processes and does them automatically so there is less human intervention with increased efficiency.
- Efficiency: ML models can analyze a huge amount of waste data and optimize critical processes. For example, they can better segregate waste by more accurately recognizing what is inside a trash bag or improve the efficiency of collection routes based on predictions of how much waste will be generated at what places and in consideration of real-time traffic conditions.
- Cost reduction: Automation and optimization trigger a colossal cost saving. Machine learning-based solutions can reduce the cost of operations by reducing manual labor, improved routing, and improved recycling operations. This makes rubbish disposal more sustainable and economical in the long run.

8.4 Applications of Machine Learning in Waste Management

Machine learning (ML) is one of the technologies that is gradually revolutionizing waste management systems. Data-driven solutions can achieve great increases in efficiency, accuracy, and sustainability. Here are some essential applications of machine learning in waste management, demonstrating how the tech is tackling nativist problems.

8.4.1 Waste sorting and classification

- **Problem statement**: Conventional waste sorting techniques often employ manual labor. Workers separate waste into paper, plastic, and metal types, but that's labor-intensive, inefficient, and prone to human error. If trash is improperly sorted, the trash could become contaminated – meaning recycling becomes less efficient.
- **Solution**: One of the ways to help with this task is to use machine learning algorithms, especially image recognition systems (e.g., convolutional neural networks (CNNs)) to automatically sort garbage. They employ visual information captured by cameras or sensors, coupled with machine learning algorithms, for effective and accurate object recognition and classification.

Example: A smart bin that has a camera attached, takes a picture of the trash that is thrown in it and uses a machine learning (ML) model to classify it as landfill, organic, or recyclable. This kind of autonomous sorting reduces human error while increasing the efficiency of waste management networks (Ramson et al., 2020).

8.4.2 Predictive analytics for waste generation

- **Problem**: There are significant differences in waste generation trends bas Waste composition is highly variable depending on seasonality, geography, and demographics. Traditional garbage collection systems struggle to keep up with these fluctuations, leading to challenges such as overflowing containers, missed pickups, and wasted resources.
- **Solution**: By analyzing historical waste volume data over a period of time, along with information on weather patterns or other demographic indicators, machine learning algorithms can assess waste generation trends. In waste management systems, it is used to optimize the collection schedule, prevent overflow, and ensure timely pickups by predicting how much waste will be generated in particular areas or neighborhoods.

Example: Taking a big city, where previously collected garbage information is processed with machine learning techniques to forecast the waste that will be produced in particular areas. It allows the municipal waste management departments to create more effective collection routes and schedules that efficiently distribute resources when it is needed (Jagtap and Rahimifard, 2019).

8.4.3 Smart routing for waste collection

- **Problem**: The inefficient route distribution for waste collection vehicle allocation results in excess fuel, labor, and time costs. By ignoring inclusion of current information, such as traffic or a filled bin, results in bypassing or slowing down the collections of such data.
- **Solution**: Controlled traffic flow, available vehicles in the vicinity, and waste quantity to be collected from households can be used by the machine learning algorithms to optimize the route of waste collection. It will allow waste management systems to formulate the best routes focused on minimum fuel consumption, time taken to cover the entire distance, and cost of operations.

Example: Waste collection trucks use machine-learned models to change their routes automatically based on real-time traffic information and bin fullness. This enables fast and efficient waste collection by cutting delays in collection and the adverse environmental impacts of waste vans (Mishra and Ray, 2020).

8.4.4 Waste reduction and recycling

- **Problem**: There are inefficiencies even with the high importance to recycling. Defective sorting, contamination, and lost chances to recover valuable materials condemn a number of recyclables to landfill.
- **Solution**: Machine learning algorithms will help to optimize the recycling process by identifying recyclable materials with a high value, automating sorting systems, as well as predicting which objects will be recycled effectively. Additionally, it can help identifying common contaminants found in recyclable materials which enable waste management companies to improve their sorting processes and lower levels of contamination.

Example: A machine learning-based environment can recognize the largest number of impurities in recycling trash, such as food on plastic or glass. By identifying these impurities, the technology assists waste management companies in perfecting their sorting operations, improving recycling efficacy while lowering the amount of waste going to landfills (Guo et al., 2021).

8.5 Case Studies of Machine Learning in Waste Management

8.5.1 Case study 1: Automated waste sorting system

(ZenRobotics, 2020)

ZenRobotics is a breakthrough in AI sorting for a sustainable future in waste management. The ZenRobotics Recycler is a unique product that combines deep learning and best-in-class robotic hardware to identify, sort, and separate recyclables from garbage, with incredible accuracy and speed. This automation has thoroughly streamlined material recycling at the waste management facility to minimize labor costs while ensuring increased yield on material separation.

8.5.1.1 System overview

ZenRobotics Recycler uses vision recognition and robotic arms to identify a wide variety of recyclable materials, including plastics, metal, and wood. It uses deep learning algorithms to interpret visual data that the cameras at the sorting station collect. These cameras inspect the waste material on a conveyor belt at all times, identifying and classifying them in real time. If detected, the goods are instantly picked up by the robot arms and moved to the right bins or containers.

8.5.1.2 Key features

- **Image recognition:** The AI system analyzes high-res images of waste, recognizing, and sorting objects into categories, such as plastic, metal, and biological materials.
- **Robotic arms:** Once the items have been identified, autonomous grippers or suction devices on robotic arms pick and place them into separate recycling bins (Devi P et al., 2024).
- **Speed and efficiency:** The machine will work 24-7 at much higher speeds than human operators and can pick as much as 2000 pieces an hour, greatly increasing throughput in waste processing operations.
- **Deep learning:** The system learns and becomes more accurate with time as it processes more data, optimizing the sorting and thus reducing errors.

8.5.1.3 Example in action

A ZenRobotics Recycler in the wild: It's deployed in waste recycling plants to sort through tons of mixed garbage. The system's cameras watch the pattern of incoming trash, and a deep learning algorithm recognizes the recyclable objects – PET bottles, aluminum cans, cardboard, etc.. AI then directs robotic arms to grab those items and put them into appropriate recycling bins.

8.5.1.4 Benefits

- **Increased accuracy:** The artificial neural network system is incredibly effective at recognizing goods and is extremely precise, which reduces the likelihood of contaminants and makes sure the recyclables are well sorted.
- **Faster processing:** Robotic arms can work at very high speeds which can significantly cut down the time taken to sort the waste compared to manual labor.
- **Cost reduction:** The automation technology can reduce labor costs and the risk of injury to workers, thereby making waste treatment more economical.
- **Sustainability:** The system guarantees proper sorting of items that could be recycled, which improves recycling rates and minimizes waste sent to landfills.

8.5.2 Case study 2: Predictive waste collection in New York City

(NYC Department of Sanitation, 2019)

As a major international city with a significant population, New York City (NYC) has many challenges with respect to its waste management system. To overcome these challenges, NYC implemented machine learning (ML) algorithms to optimize routes and make real-time predictions on waste pickup. The idea was to enhance efficiency of waste collection, minimize fuel consumption, ensure timely waste collection, and lower operating costs.

8.5.2.1 System overview

The DSNY (New York City Department of Sanitation) collaborated with machine learning experts to create a predictive garbage collection system that utilizes real-time data to optimize garbage truck routes. By analyzing the historical waste disposal data, as well as various parameters like nearby population, traffic conditions etc., the system identifies the amount of waste produced at individual waste disposal points and figures out the best routes to be taken by waste disposal vehicles.

8.5.2.2 Key features

- **Predictive analytics:** ML models use information such as previous wastage generation patterns, seasonal patterns, and time of day profiles to predict the quantity of wastage in each region.
- **Optimization of collection routes:** The algorithms use these predictive data to build the best collection routes, in order to guarantee that trucks drive through the shortest lines and avoid congested roads.
- **Real-time data integration:** The technology deploys state-of-the-art data, including real-time information about traffic or vehicle availability, to change roads if needed.
- **Dynamic scheduling:** Technology enables adjustments to collection schedules in real-time, preventing missed pickups, overflows in the rolling stock, and resource underutilization.

8.5.2.3 Example in action

The system typically analyzes data sensing from trash cans and past patterns of trash collecting. The machine learning algorithm predicts that a specific neighborhood discards more rubbish on a certain day. It then configures the collection route for garbage trucks, factoring in traffic conditions, the expected amount of waste in particular areas, and the location of waste bins. As a result, trucks can reroute on the spot, therefore no waste goes uncollected, and no time gets wasted.

8.5.2.4 Benefits

- **Fuel savings:** By calculating optimal routes for the garbage trucks, supported by our system, the travel distance of the garbage trucks will be shortened, thereby saving fuel and creating a smaller carbon imprint associated with waste collection.
- **More accurate pickups**: With better estimates and improved routes, waste collection staff can ensure services are completed, thereby reducing the likelihood of missed collections.

  This allowed for cost-effective operational management, as the waste routes are connected to each other, preventing the need to send the trucks back and forth on the same route. The data tracks the people's habits and actions for housing.

- **Environmental impact:** The technology reduces emissions by optimizing routes and reducing fuel consumption, leading to greener waste collection in NYC.

8.6 Challenges in Implementing Machine Learning for Waste Management

Though machine learning (ML) can make a big difference in waste management, there are several obstacles that need to be addressed before such technologies can be utilized effectively. These issues vary from data-related challenges to economic and social constraints. The following section presents some of the most critical issues in deploying machine learning systems for garbage management (Joshi et al., 2021 & Schneider and Turner, 2019).

8.6.1 Data availability and quality

Challenge: To be efficient, machine learning algorithms require massive amounts of quality and well-annotated datasets. Such data is utilized in waste management to define the kinds of rubbish produced, waste composition, and collection trends. Yet collecting and structuring such information proves to be a big challenge.

- **Data messiness:** Waste information is often disorganized, inconsistent, and incomplete. For instance, there could be gaps in data on the types of materials recycled or there are discrepancies in how trash is categorized across regions.
- **Lack of proper labels:** Labeling waste information (i.e., classifying waste as recyclable, organic, or not recyclable) is a time and money-intensive process. In addition, if labeling is not specific, training machine learning models is harder, which could result in suboptimal model performance.

Impact: Poor data quality and inadequate datasets make it difficult to train machine learning models, resulting in inaccurate predictions and poor waste management procedures.

8.6.2 High initial investment

Challenge: Implementing machine learning-based waste management systems needs a considerable financial investment. This covers the expense of:

- **Sensors and cameras:** Cameras, infrared sensors, and weight sensors are essential for data collecting, especially in automated garbage sorting systems.
- **Robotic systems:** The deployment of robots (for example, for sorting or rubbish collecting) entails significant upfront costs for procuring and maintaining these technologies.
- **ML infrastructure:** Building the infrastructure for ML model training, data storage, and processing can be expensive, necessitating investments in strong hardware and software platforms.

Impact: Many waste management organizations, particularly those in underdeveloped countries or with limited finances, may find the high cost of deploying these technologies prohibitive. As a result, these organizations may be unable to use machine learning systems, resulting in inefficiencies in waste management.

8.6.3 Resistance to change

Challenge: Workers in traditional areas, such as trash management, frequently object to the introduction of machine learning and AI technologies. Several factors contribute to this resistance.

- **Fear of job loss:** Workers may be concerned that automation, robotics, and machine learning systems may replace human workers, causing unemployment in the business.
- **Disruption of established processes:** Historically, waste management relied on manual labor and set procedures. The introduction of new technology has the potential to disrupt traditional processes and compel workers to learn new skills, which may elicit opposition.
- **Cultural resistance:** Employees who have used traditional methods may be suspicious about the usefulness and dependability of AI-powered systems, perceiving them as unreliable or superfluous.

Impact: Resistance to change can impede the adoption of machine learning technology, preventing firms from reaping the full benefits of waste management automation and optimization.

8.6.4 Regulatory and ethical concerns

Challenge: The use of machine learning in waste management raises a number of regulatory and ethical concerns that must be carefully addressed:

- **Data privacy:** The use of sensors and cameras for waste sorting, waste bin monitoring, and waste truck tracking can pose data privacy concerns, particularly if personal information (such as location data) is acquired without authorization.
- **Fairness and equity:** An AI system for waste management must be constructed in such a way that decision-making is fair and equitable. Automated garbage sorting systems, for example, must avoid disproportionately affecting specific demographic groups or waste streams.
- **Accountability and transparency:** Machine learning algorithms often work as "black boxes," and therefore transparency is needed regarding how these systems arrive at decisions. This is especially relevant in public sectors such as garbage collection where citizens could be skeptical of automated systems' neutrality.

Impact: Regulators and waste firms must ensure that use of machine learning technologies does not breach privacy laws and ethical guidelines. Overlooking these challenges primarily results in public outcry, legal action, and loss of trust in the system.

8.7 The Future of Machine Learning in Waste Management

8.7.1 Smart cities and waste management

The next step toward smart cities will also be to ensure that waste management systems are capable of developing into the most efficient, transparent, and sustainable systems of waste disposal, which is where machine learning is going to be extremely crucial. Smart waste and creating a circular economy – smart waste management will involve embedded sensors in waste bins, AI-driven sorting plants and real-time optimization software which will revolutionize the waste disposal in cities.

8.7.2 Integration with a circular economy

If it is integrated with a circular economy designs, machine learning could improve the respective lifetime of commodities and waste. AI systems could predict which forms of reuse, recycle, or reduce would be optimal improving sustainability goals.

8.7.3 Ethical AI in waste management

With machine learning systems embedded in waste management solutions, ethical considerations of AI will need to take precedence. How can long-term trust and success hinge on ensuring these systems are transparent, equitable, and privacy-respecting.

8.8 Conclusion

Machine learning will be a transformative tool for leverage in waste management, increasing efficiency, sustainability, and profit. By automating operations like sorting, pickup optimization, and waste forecasting, machine learning can completely change garbage disposal systems to be more efficient in terms of process and cost. In the future, smart systems will increasingly interact with other industries such as urban planning, energy, and environmental protection in terms of waste management, promoting a more holistic and sustainable solution.

In the future, we look forward to more intelligent solutions which not only improve waste management but also contribute to wider environmental goals. To fully realize these benefits, however, issues such as data quality, investment costs, and ethics need to be addressed. The last but not least, application of machine learning will be in constructing smarter, greener cities, and waste management system.

References

Gao, Y., Wang, J. and Xu, X. (2024). Machine learning in construction and demolition waste management: Progress, challenges, and future directions. Automation in Construction, 162: 105380.

Guo, H., Wu, S., Tian, Y., Zhang, J. and Liu, H. (2021). Application of machine learning methods for the prediction of organic solid waste treatment and recycling processes: A review. Bioresource Technology, 319: 114-127.

Jagtap, S. and Rahimifard, S. (2019). The digitisation of food manufacturing to reduce waste – Case study of a ready meal factory. Waste Management, 87: 387-397.

Joshi, L. M., Bharti, R. K. and Singh, R. (2021). Internet of things and machine learning-based approaches inthe urban solid waste management: Trends, challenges, andfuture directions. Expert Systems, 39(5): 1-23.

Mishra, A. and Ray, A. K. (2020). IoT cloud-based cyber-physical system for efficient solid waste management in smart cities: a novel cost function based route optimisation technique for waste collection vehicles using dustbin sensors and real-time road traffic informatics. IET Cyber-Physical Systems: Theory & Applications, 5(4): 330-341.

NYC Department of Sanitation (2019). "Smart Waste Management: The Role of Machine Learning in Optimizing Waste Collection in New York City." Retrieved from: https://www.nyc.gov.

Ramson, J., Suresh, V., Moni, J. and Anagnostopoulos, T. T. (2020). An IoT-based bin level monitoring system for solid waste management. Journal of Material Cycles and Waste Management, 23(3).

Recycling waste classification using optimized convolutional neural network. Available at https://www.sciencedirect.com/science/article/abs/pii/S0921344920304493.

Schneider, J. and Turner, M. (2019). Data Quality in Smart Waste Management Systems: Challenges and Solutions. Waste Management & Research, 37(6): 532-539.

Smart waste management: A paradigm shift enabled by artificial intelligence. Available at https://www.sciencedirect.com/science/article/pii/S2949750724000385.

Arora, M., Ahmad, V., Kumar, R., & Singh, R. (2025). The Impact of Industry 4.0 Technologies on Operational Excellence Methods: Application of Fuzzy Topsis. Engineering Management Journal, 1-12.

Ahmad, V., Arora, M., Sayal, A., & Kumar, R. Exploring Trends in E-Business and Sustainable Finance: An Integrated Analytical Approach. Frontiers in Sustainable Cities, 7, 1549874.

Devi, P., Desai, S. S., Kumar, R., Naruka, M. S., & Tripathi, V. (2024, December). A Strategic Innovations for Stock Market Optimization by Using Data Science & Explainable AI in High Frequency Trading. In 2024 International Conference on IoT Based Control Networks and Intelligent Systems (ICICNIS) (pp. 1510-1515). IEEE.

Kumar, R., Goel, R., Singh, T., Mohanty, S. M., Gupta, D., Alkhayyat, A., & Khanna, R. (2024). Sustainable Finance Factors in Indian Economy: Analysis on Policy of Climate Change and Energy Sector. Fluctuation & Noise Letters, 23(2).

Kumar, R., Kandpal, B., & Ahmad, V. (2023). Industrial IoT (IIOT): Security threats and countermeasures. *2023 International Conference on Innovative Data Communication Technologies and Application (ICIDCA)*, 829–833. IEEE

What Is Unsupervised Learning? Available at https://eastgate-software.com/what-is-unsupervised-learning/.

ZenRobotics (2020). "ZenRobotics Recycler: AI-Based Robotic Waste Sorting." Retrieved from: https://www.zenrobotics.com.

9

Bridging Corporate Social Responsibility and a Circular Economy in Manufacturing in Light of Economic, Technological, and Policy Barriers: Addressing Integration Challenges and Future Directions

Arpana Kumari[1], Richa Goel[2,3], and Rupa Khanna[3]

[1]Symbiosis Centre for Management Studies, Symbiosis International Deemed University, India
[2]Graphic Era University, India
Symbiosis Centre for Management Studies,
Symbiosis International Deemed University, India
[3]Graphic Era University, Dehradun
E-mail: arpanasngh15@gmail.com; richasgoel@gmail.com; dr.rupakhanna@gmail.com

Abstract

The corporate social responsibility (CSR) and a circular economy (CE) practices are prerequisites for the promotion of sustainability in the manufacturing sector. While CSR ensures ethical and responsible business, CE emphasizes efficiency in the use of resources, minimization of waste, and closed-loop production. Despite their common goals of sustainability, the integrated role of CSR and CE in manufacturing needs to be explored to uncover their potential in larger phenomena. This chapter discusses the current literature on CSR and CE, highlights key research gaps, and presents directions for future research. There are some specific challenges that hamper the integration between CSR and CE practices and are essential to be reviewed. This chapter conducted a systematic review of literature to explore these barriers, which

DOI: 10.1201/9788743806905-9

are weak economic and technological support, regulatory and policy hurdles, and poor stakeholder engagement. It is important to closing these gaps which calls for research in this area. The research significantly contributes to the need of strategies that can integrate the practices of CSR and CE to achieve more sustainable manufacturing industry.

Keywords: Corporate social responsibility, circular economy, manufacturing industry, sustainability, stakeholder engagement, policy barriers.

9.1 Introduction

The manufacturing sector is considered a highly contributing sector in the economic growth of any country, specifically India. This sector generates a large number of job opportunities, expands external and internal trade and business landscape. Its transformative growth has also led to many opportunities in other sectors and addressed social and environmental concerns of the country. The advancement in small, medium, and large-scale manufacturing units has created independence of products in India, which has minimized its dependency for the product on other countries. However, this sector is facing some key challenges such as waste creation, global warming, and pollution. To overcome these issues, organizations are adopting corporate social responsibility (CSR) and a circular economy (CE) practices following the concept of sustainability.

CSR is a practice followed in corporations to run their business operations based on ethics, and contribute to societal benefits. It incorporates a wide range of practices that include environmental preservation, employment rights, fair labor practices, supply chain sustainability, and corporate governance. CSR has become widely accepted as an integral part of corporate strategy designed to integrate financial performance with societal and environmental factors. CE, on the contrary, is a systemic process of economic growth aimed at minimizing waste and fostering the recycling of resources over and over. CE differs from the conventional linear economy, based on a "take-make-dispose" scenario, as it focuses on closed-loop production, efficiency in the use of resources, recycling, and remanufacturing (Saidi & Radzi, 2023). Although CSR and CE have common goals of sustainability and responsible business, their integration is still limited in the manufacturing industry (Berkel, 2023). Most companies continue to view CSR as a distinct corporate function instead of integrating it into core business strategies that foster circularity. It is thus important to understand how CSR can promote the adoption of CE in order to push sustainable industrial development forward.

The manufacturing sector involves enormous quantities of raw materials and energy and produces large amounts of waste. This has resulted in increasing regulatory, consumer, and investor pressure on companies to become more sustainable. CE principles are compelling pathways to resolving these issues by designing waste out of products, lengthening product lifespans, and ensuring responsible consumption and production (Ellen MacArthur Foundation, 2022; De Angelis et al., 2018). Yet, the change to a circular system involves thoughtful shifts in business models, supply chains, and ways of operating.

CSR can be a mechanism for the adoption of CE through integrating sustainability values within business decision-making frameworks. Corporates with a greater focus on CSR projects tend to spend more on adopting environmentally friendly manufacturing practices, sustainable energy alternatives, and high-technology recycling mechanisms. Further, CSR drives stakeholder collaboration toward businesses, governments, customers, and the supply chain ecosystem for designing circular products and solutions. In spite of these possible integrations, aligning CSR with CE in manufacturing is proving challenging. CSR is often seen by most firms as a compliance-based activity instead of a central part of the business strategy. Financial and technological limitations also usually hamper the adoption of circular practices. There is also a necessity to conduct more research on how CSR can be suitably utilized for driving CE efforts across various manufacturing industries.

This chapter seeks to critically analyze the association between CSR and CE in the manufacturing sector, determine major research gaps, and suggest directions for future research. The research objectives of this chapter are as below:

1. To examine economic, technological, and policy barriers to CE adoption
2. To assess stakeholder engagement mechanisms that support CE transition
3. To outline future research avenues to further CSR-driven CE practice

This research offers advancing knowledge on CSR and CE integration and contributes to academic and business literature on sustainable manufacturing. Through the investigation of how CSR can propel CE adoption, this research offers insight into what policymakers, industry executives, and researchers want to form complete sustainability approaches. Additionally, the discovery of shortcomings in previous research will inform subsequent studies for advancing CSR and CE principle integration.

9.2 Review of Literature

9.2.1 Economic, technological, and policy barriers to CE adoption

One of the significant gaps in the literature is the lack of comprehensive policy frameworks that can facilitate the integration of CSR and CE in Asian manufacturing industries. While some studies emphasize the role of government interventions and regulations in promoting CE practices, there is limited research on how these policies can be designed and implemented effectively across diverse Asian countries (Saidi & Radzi, 2023) (Anbumozhi & Kim, 2016). Due to limited research on the integrated role of CSR and CE, it is challenging to understand their combined role and benefits to the organization and society at large. Considering the significance of both concept organisations are CE and CSR Research is required to develop tailored policy recommendations that align with the unique challenges and opportunities of each country or region.

9.2.1.1 Economic barriers

In the manufacturing sector, the adoption of a circular economy (CE) is understood to come with economic challenges that limit its implementation. These challenges are caused by capital issues, low-grade technology, and policy and regulatory issues, which altogether make it difficult to transition to sustainable practices. It is essential to view these challenges in specifying the policies that encourage the adoption of CE. Moreover, another challenge is the lack of government support, which adds to companies' limitations for adopting CE and CSR together, as there is no provision of receiving financial incentives that could enable CE and CSR integration (Holly et al., 2023). Our research aims to explore the possibilities where the successful integration of CSR into business with CE can bridge this gap. Organizations that link their CSR efforts to CE practices tend to attract impact investors and enjoy favorable government policies (Ferreira et al., 2021). Additionally, CSR-oriented sustainability reporting promotes the reputation of corporations, which, in turn, generates higher stakeholder trust and customer loyalty (Du & Vieira, 2012).

Changing to CE can involve large amounts of initial capital for new processes and technology (Rao et al., 2024). Insufficient government support and insufficient monetary incentives or support from governments are common complaints among many manufacturers, which is instrumental in making the change (Holly et al., 2023). Regulatory complications such as cumbersome regulations working with the regulations may be difficult, with diverse standards across the globe that make it difficult to comply (Salman et al., 2024). Market demand, influenced by varying customer tastes, can discourage manufacturers from making investments in CE, since the return on investment would be indeterminate (Holly et al., 2023). Although these are powerful challenges, they also underscore the importance of cooperative action among stakeholders, such as policymakers, to develop enabling frameworks that allow for a successful transition to a circular economy. Others, however, state that the economic advantages of CE, like job creation and improved resource efficiency, might outweigh these hurdles, presuming that an emphasis on long-term benefits would motivate more firms to embrace circularity (Mohamed, 2024). Also, CSR-based sustainability reporting fuels company reputation, generating more stakeholder trust and customer loyalty (Du & Vieira, 2012). CSR can also be crucial in solving these regulatory problems. Through proactively participating in multi-stakeholder partnerships, companies are able to shape policy structures that ensure CE-supportive regulation, further fostering a sustainable environment for transitions (De los Rios & Charnley, 2017). Companies that incorporate CSR into CE strategies put themselves at an advantage for competitive success through boosting brand value, building innovation, and ensuring long-term profitability (Goyal et al., 2023). Thus, a collective effort by policymakers, companies, and other stakeholders is necessary to creating enabling frameworks that promote CE adoption while strengthening CSR commitments.

9.2.1.2 Technological barriers

The integration of emerging technologies, including Industry 4.0 and digital innovation, is essential for the effective deployment of CE practices. Integration of digital technology with adoption of digital technologies like IoT, AI, and digital twins is essential to monitor resource flows and enhance product life cycles. Data security and privacy, which includes the threat of data breaches and protection of customer data, are key concerns. This is specifically important in manufacturing sector where data export is common (Bag et al., 2021). Moreover, the practices of standardization in products and processes also hinder the CE adoption (Rao et al., 2024). Operational challenges include remanufacturing experiences, uncertainties regarding core quality, return times, and variations in processes. These challenges complicate the integration of remanufacturing into CE frameworks (Kim et al., 2024). Further in advanced technologies, while machine learning and data-driven models offer solutions, their implementation is complex and requires significant technological advancements (Kim et al., 2024). However, the lack of empirical evidence and the nascent stage of research on the synergy between these technologies and CE pose significant challenges (Denu et al., 2023). There is limited research on the technological and innovation barriers faced by Asian manufacturing industries in integrating CSR and CE practices (Piyathanavong et al., 2024) (Chen, 2024).

9.2.1.3 Policy barrier

Lack of well-established and comprehensive policies handicaps the smooth shift to CE practices. Weakness in enforcement of current environmental laws promotes noncompliance, which denies CE efforts (Chhimwal et al., 2022). Poor economic incentives in the form of tax relief or subsidies discourage manufacturers from promoting sustainable practices (Nandha Gopan & Balaji, 2023), which also brought forth the issue of quality checks. Also, the absence of harmonized regulations makes CE practices confused and inconsistent to enforce across the sector (Zaidi & Chandra, 2024), wherein standards and regulatory barriers" "strategic barriers and "supply chain management and technology are specialized barriers when deliberated into clothing industry.

9.2.2 Stakeholder engagement mechanisms that support CE and CSR transition

Engagement and collaboration of stakeholders, including governments, companies, and civil society, are crucial for the implementation of effective CSR and CE practices (Vasilev & Nikolova-Alexsieva, 2024) (Sugimoto, 2016). The development of effective stakeholder engagement and cooperation for fostering CSR and CE practices in Asian manufacturing sectors should be the focus of research. Stakeholder engagement and collaboration are key to the effective implementation of a circular economy (CE) and corporate social responsibility (CSR) policies. Governments, companies, and civil society need to collaborate to drive sustainable changes in the Asian manufacturing sector (Vasilev & Nikolova-Alexsieva, 2024; Sugimoto, 2016). Research indicates the need to develop organized frameworks that enable multi-stakeholder collaboration, with smooth integration of CSR and CE principles across industries.

Table 9.1 Previous studies

A circular economy strategies for resource efficiency and sustainable development in manufacturing industries	Rao, Y., JP, R., Thomas, S., Agrawal, D., S.M, U. D., & Hajra, V. (2024)	Analyze a circular economy strategies that improve resource efficiency and contribute to sustainable development in manufacturing	Highlights the importance of closed-loop supply chains, waste reduction, and technological interventions in CE adoption	Limited to South Asian context; lacks empirical validation through case studies	Rao et al., 2024
Challenges on the way to a circular economy from the perspective of the Austrian manufacturing industry	Holly, F., Kolar, G., Berger, M., Fink, S., Ogonowski, P., & Schlund, S. (2023)	Identify barriers faced by Austrian manufacturers in transitioning toward a circular economy	Policy inconsistencies, high initial investment, and lack of market demand hinder CE adoption	Country-specific insights; findings may not generalize across regions	Holly et al., 2023
The pathways to lean manufacturing for a circular economy: Implications for sustainable development goals	Salman, S., Taqi, H. Md. M., Nur, S., Awan, U., & Ali, S. M. (2024)	Examine how lean manufacturing principles align with a circular economy objectives and SDGs	Identifies synergy between lean manufacturing and circularity in reducing waste and increasing efficiency	Focuses on lean without integrating digitalization aspects	Salman et al., 2024
Challenges in the implementation of a circular economy in manufacturing industry	Chhimwal, M., Agrawal, S., & Kumar, G. (2022)	Explore the barriers hindering CE implementation in manufacturing firms	Identifies financial, technological, and policy barriers affecting CE adoption	The study lacks empirical validation with industry-specific data	Chhimwal et al., 2022

Table 9.1 *Continued.*

Indian automotive supply chains: barriers to a circular economy for sustainable development	Nandha Gopan, S. & Balaji, M. (2023)	Investigate barriers to CE in the Indian automotive sector	Finds lack of infrastructure, regulatory gaps, and supply chain inefficiencies as key hurdles	Limited to Indian automotive industry; does not include cross-sectoral analysis	Nandha Gopan & Balaji, 2023
The challenges to a circular economy in the Indian apparel industry: a qualitative study	Zaidi, A.A. & Chandra, R. (2024)	Examine barriers to CE in the Indian apparel industry	Highlights lack of consumer awareness, low investment in CE practices, and regulatory bottlenecks	Study focuses only on qualitative aspects without quantitative validation	Zaidi & Chandra, 2024
The trend of a circular economy studies in Asian countries: A bibliometric analysis	Saidi, N. A. & Radzi, N. A. (2023)	Conduct a bibliometric review of CE research in Asia	Identifies key research clusters and gaps in CE implementation in Asia	Focuses on publication trends rather than practical applications	Saidi & Radzi, 2023
Stakeholder engagement and environmental practice adoption: The mediating role of process management practices	Jakhar, S. K. (2017)	Investigate the role of stakeholder engagement in environmental practice adoption	Stakeholder involvement enhances environmental sustainability initiatives in manufacturing	Lacks empirical analysis beyond the manufacturing sector	Jakhar, 2017
Industry 4.0 deployment for a circular economy performance –Understanding the role of green procurement and remanufacturing activities	Sahoo, S., & Jakhar, S. K. (2024)	Analyze the impact of Industry 4.0 technologies on a circular economy performance	Green procurement and remanufacturing play a crucial role in achieving circularity	Requires more case studies from diverse manufacturing sectors	Sahoo & Jakhar, 2024

Table 9.1 *Continued.*

Moving toward a circular economy in manufacturing organizations: the role of circular stakeholder engagement practices	Fobbe, L., & Hilletofth, P. (2023)	Examine how stakeholder engagement can facilitate CE transitions in manufacturing firms	Effective stakeholder engagement accelerates CE adoption through collaborative networks	Limited focus on digital tools for stakeholder engagement	Fobbe & Hilletofth, 2023
Business sustainability: It is about time	Bansal, P., & DesJardine, M. R.	To explore the temporal dimensions of business sustainability and the challenges in integrating long-term sustainability goals into business strategies	The study emphasizes that sustainability should be viewed with a long-term perspective and not just as short-term gains. It highlights the role of institutional pressures in shaping sustainability practices	The research primarily focuses on conceptual discussions without empirical validation, limiting its generalizability	Bansal & DesJardine, 2014
Product design and business model strategies for a circular economy	Bocken, N. M. P., de Pauw, I., Bakker, C., & van der Grinten, B.	To investigate how product design and business models can be structured to support the circular economy	Proposes several strategies, including design for longevity, reuse, remanufacturing, and recycling to facilitate a circular economy transitions	Limited empirical data to validate the proposed strategies in diverse industrial sectors	Bocken et al., 2017
Skills and capabilities for a sustainable and a circular economy	De los Rios, I. C., & Charnley, F. J.	To examine the role of design skills and capabilities in promoting sustainability and a circular economy principles	Identifies key skills such as systems thinking, material innovation, and cross-disciplinary collaboration as essential for a circular economy adoption	Focuses on the design sector, limiting its applicability to other industries	De los Rios & Charnley, 2017

Table 9.1 *Continued.*

Striving for legitimacy through corporate social responsibility: Insights from oil companies	Du, S., & Vieira, E. T.	To explore how oil companies use CSR initiatives to gain legitimacy	Highlights that CSR efforts enhance corporate reputation and stakeholder trust but often serve strategic rather than ethical purposes	The study is limited to the oil industry, restricting its generalizability to other sectors	Du & Vieira, 2012
Sustainability, a circular economy, and corporate social responsibility: Mapping the interconnections	Ferreira, F. A., Jalali, M. S., & Ferreira, J. J.	To analyze the interconnections between sustainability, the circular economy, and CSR	Finds that CSR can act as a bridge to integrate a circular economy practices within corporate sustainability agendas	Conceptual in nature, requiring empirical testing	Ferreira et al., 2021
Corporate social responsibility and sustainable business practices: A strategic perspective	Goyal, S., Rahman, Z., & Kazmi, A. A.	To evaluate how CSR influences sustainable business practices	Demonstrates that CSR initiatives lead to improved sustainability outcomes and competitive advantages	The study does not address industry-specific variations in CSR implementation	Goyal et al., 2023
Financial barriers to a circular economy implementation in the manufacturing sector	Holly, T., Smith, J., & Williams, R.	To identify financial constraints hindering a circular economy adoption	High upfront costs and uncertain return on investment are major barriers to CE transition	Lacks case studies to provide practical insights	Holly et al., 2023

Table 9.1 *Continued.*

Long-term economic benefits of a circular economy adoption in manufacturing	Mohamed, K.	To assess the financial viability of a circular economy strategies in manufacturing	Finds that although initial investment costs are high, long-term benefits such as cost savings and resource efficiency outweigh them	Limited to a case study approach, requiring broader validation	Mohamed, 2024
Economic challenges and technological gaps in a circular economy adoption: A critical review	Rao, P., Kumar, S., & Patel, R.	To examine economic and technological barriers to a circular economy adoption	Highlights the need for government incentives and technological standardization to support CE implementation	Does not provide specific industry-focused recommendations.	Rao et al., 2024
Regulatory complexities in the transition to a circular economy: A cross-sectoral analysis	Salman, F., Brown, L., & White, M.	To analyze regulatory challenges affecting a circular economy adoption across industries	Finds that inconsistent policies and weak enforcement mechanisms hinder effective CE implementation	Primarily focuses on developed economies, lacking insights into emerging markets	Salman et al., 2024

Stakeholder interaction has a key role in adopting environmental practices within the Indian manufacturing industry, defining sustainability marketing practices and supplier relationships (Jakhar, 2017). Excellent supplier relationships do not only help in improving efficiency but also support the competitiveness of manufacturing companies (Kumar & Kazmi, 2016). Moreover, responsive manufacturing practices – defined as frequent interaction with suppliers, customers, and regulators – have been reported to push the performance indices to improve efficiency and customer satisfaction.

Shifting toward CE in Indian manufacturing involves stakeholder engagement mechanisms that are clearly defined. Past studies show that rigorous training programs raise the awareness of employees regarding CE principles, leading to an organizational culture based on sustainability (Singh et al., 2025). In addition, strategic partnerships among companies, suppliers, and consumers enable the co-creation of circular solutions, including remanufacturing schemes and sustainable purchasing (Sahoo & Jakhar, 2024). Cross-sectoral collaboration – such as partnerships with government agencies, NGOs, and academia –allows CE adoption to leverage varied expertise and viewpoints, enhancing policy formulation and real-world implementation (Fobbe & Hilletofth, 2023). Through the inclusion of these engagement mechanisms, Indian manufacturing companies can speed up their shift to a circular economy while strengthening CSR commitments.

9.3 Discussion and Future Research Avenues

This chapter offers significant contribution in the literature of CE and CSR adoption. The chapter reviews key literature and identified the specific barriers of CSR and CE in manufacturing sector. The chapter also highlights the points to overcome these barriers. The chapter discussed about the technological barriers, economical and policy-related barriers to CSR and CE adoption in integrated from. It is highlighted that stakeholder engagement is also an important factor which to drive the CE and CSR-driven sustainable path (Holly et al., 2023; Rao et al., 2024). The most prominent issue of CE adoption is the cost of conversion to the circular form as right now companies are following linear form of manufacturing process. CE adoption involves a large scale of capital cost where regulatory support of required level is missing which can be detrimental to organizations when market is unstable (Norhayati Nordin et al., 2023). Moreover, currently the government has also not shown any major initiatives in terms of tax benefits and subsidies for CE adoption (Sarah R. Prendeville et al., 2023).

CSR plays a role in surmounting such economic challenges by aligning economic incentives with sustainable goals in the long run. Companies that incorporate CSR into their business models can leverage sustainability reporting and value-based financial systems to serve accountable investors and attract green financing (Goyal et al., 2023). Through the application of CSR-based approaches, firms can communicate the cost savings over the long term concerning resource efficiency, reduction of waste, and power efficiency to build investor and consumer trust in CE-style investments (Mohamed, 2024). In addition, companies involved in CSR activities, like ethical purchasing and sustainable sourcing, establish a system in which circular principles are incorporated into corporate governance. CSR activities that minimize environmental impacts and encourage responsible production can also strengthen brand image and customer loyalty, ultimately influencing market demand for CE-oriented goods and services (Du & Vieira, 2012). These economic advantages provide evidence of how companies are able to leverage sustainability challenges into competitive advantages. Another essential stumbling block to implementing CE is technological integration. Various sectors of industries fail to use innovative digital technology like the Internet of Things (IoT), artificial intelligence (AI), and digital twins necessary for tracking resources and automating processes (Bag et al., 2021). A lack of standardized digital infrastructure is the cause of inefficiencies within waste management and remanufacturing and of operational unsustainability (Kim et al., 2024). Operational barriers, such as inconsistent product quality and return lead times, further hinder the setup of circular supply chains.

The initiatives of CSR will fill in those technology gaps and support corporate investment in sustainable innovation and digital transition. Firms in organizations that value CSR tend to invest in research and development (R&D) on green technology, smart manufacturing, and solutions for waste minimization (Goyal et al., 2023). By integrating CSR into their technology adoption approach, companies can develop scalable models of remanufacturing and recycling, making CE activities both environmentally friendly and economically sound. In addition, industry partnerships driven by CSR enable knowledge sharing and technology transfer among companies, policy-making institutions, and research organizations. Collaboration and joint ventures among firms, suppliers, and universities can improve technological abilities through co-creation of CE solutions, including sustainable material design innovations and digital tracking systems for resource optimization (Sahoo & Jakhar, 2024). These collaborations are consistent with stakeholder engagement principles, further supporting the role of CSR in CE implementation.

Weak regulation and enforcement measures are another considerable challenge to implementing CE. Across most Asian economies, decentralized environmental regulations contribute to uncertainty, complicating compliance for producers (Chhimwal et al., 2022). Additionally, policy models do not readily support the socio-economic and cultural determinants governing CE adoption within various sectors (Saidi & Radzi, 2023). Unless there are effectively defined and cohesive regulations, manufacturers cannot bring operations in line with sustainability objectives and, therefore, have low levels of CE adoption.

CSR can transform the role it plays in addressing policy impediments by pushing for sustainable regulations as well as cross-industry adherence. CSR-seeking firms undertake sustainability reporting and corporation disclosures, urging transparency and accountability in environmental governance (Salman et al., 2024). CSR-seeking firms may influence the policy-making process when they are proactively involved in policy debates, public-private forums, and joint partnerships. That way, these firms can create regulatory frameworks accommodating CE uptake coupled with ensuring their policies meet their industry requirements.

Moreover, CSR-oriented companies tend to lead in establishing voluntary environmental standards that exceed regulatory requirements. Through best sustainability practices in manufacturing, labor ethics, and green product design, companies can set standards that would prompt broader industry application of CE principles (Singh et al., 2025). Stakeholder engagement is the significant element in the adoption of CE, especially as per the research paper, CSR can be one of the major enablers with its features of social, business, and academia integration which involves business, government, and society (Vasilev & Nikolova-Alexsieva, 2024). Effective stakeholder engagement is a mechanism that ensures CE practices are framed in collaboration through partnerships of multiple sectors, thus ensuring sustainability transformations

One of the notable contributions of this research is exploring the interface of CE adoption and business performance. Although there can be an upfront cost, some of the long-term benefits that come out of CE include profit maximization, improved reputation for the company brand, and business differentiation (Sahoo & Jakhar, 2024). Companies incorporating CSR into business models have enhanced financial performance due to the correlation between sustainability targets and market forces in demand for environmental-friendly goods and services.

9.3.1 Future research avenues

Future studies must investigate the formulation of region-specific policy designs that respond to economic and regulatory differences in various manufacturing industries. Best practices from CE leaders across the world can help identify efficient policy mechanisms that are specifically suited for Asian markets (Saidi & Radzi, 2023). Economic viability and financial incentives should be part of further research, especially into the efficacy of financial models, such as subsidies, tax breaks, and public-private partnerships, in stimulating CE investments. This research should also evaluate the long-term fiscal advantages of CE, for example, cost savings, creation of employment, and efficiency in using resources (Mohamed, 2024).

Empirical research can be taken ahead in to technology integration and digital innovation to explore the ways in which new technologies like AI, IoT, blockchain, and big data analytics can be leveraged to promote CE adoption. Research needs to find out how technological barriers are to be bridged by interoperability and standardized digital paradigms (Bag et al., 2021; Kim et al., 2024). More empirical studies through sector-based case studies are needed to test theoretical frameworks and yield actionable findings. Research must examine the implementation of CE in different sectors like automotive, apparel, and electronics to determine sector-specific challenges as well as best practices (Zaidi & Chandra, 2024; Nandha Gopan & Balaji, 2023). Stakeholder engagement and cooperation mechanisms for developing systematic models of engagement need to be explored so that there will be more coordination among companies, governments, and civil society actors for the adoption of CE. Examining the contributions of cross-sector collaboration, corporate social responsibility (CSR) campaigns, and public awareness marketing can be a framework for effective stakeholder cooperation (Jakhar, 2017; Fobbe & Hilletofth, 2023). Consumer awareness and market demand: It must examine how consumer awareness would enhance CE uptake and determine ways to reinforce demand for eco-friendly products. Studies in consumer psychology and behavioral economics can guide organizations to design marketing strategies that will promote circular consumption behaviors (Zaidi & Chandra, 2024).

Research needs to examine the impacts of CE adoption on firm performance, such as profitability, competitiveness, and brand image. Long-term studies of companies that have made the shift to CE can reveal useful insights into advantages and constraints of circular business models (Sahoo & Jakhar, 2024). By addressing these research avenues, scholars and policymakers can develop more effective strategies to overcome the barriers to CE adoption and facilitate a sustainable transition in manufacturing industries.

This research contributes greatly to theory through the synthesis of circular economy (CE) and corporate social responsibility (CSR) paradigms, especially for the Indian manufacturing industry. This research adds value to the literature through the display of how sustainability initiatives under CSR can overcome limitations in adopting CE. While extant literature has tended to discuss CE mainly through economic and legal lenses, this research takes it a step further by highlighting CSR as a facilitator of sustainable transitions.

9.3.2 Theoretical implication

The research emphasizes the significance of involving economic, social, and environmental stakeholders in the implementation of CE. Governments, businesses, civil society, and academia are essential in influencing policies and business strategies that enhance circularity. The study supports the notion that collaboration among these stakeholders can enable the adoption of CE through the encouragement of responsible business practices, enhancing sustainability marketing approaches, and improving supplier relationships.

Further, the research adds to the theory on CE uptake. Inefficient enforcement and diversified regulations are sustainability barriers, especially in Indian manufacturing. CSR-led policy lobbying and business actions, however, can reinforce regulatory voids, leading to better alignment with environmental regulations. Through institutional policy matching of CE initiatives, firms can overcome regulatory hurdles while fostering sustainability.

The second theoretical contribution of this research is the development of sustainable business model innovation. The study demonstrates how companies can utilize CSR activities to incorporate circularity into their business models. Investments in green technologies, digitalization, and moral supply chain management allow companies to create sustainable business practices. Through the emphasis of these strategies, the research presents a model for companies to align economic growth with environmental stewardship.

9.3.3 Managerial implications

From a management point of view, this research provides management recommendations to business executives, policymakers, and industry players interested in promoting CE uptake through CSR-based approaches. The results highlight the importance of ensuring that business CSR initiatives are aligned with CE goals to establish a responsible and sustainable business climate. By integrating sustainability values into business operations, corporations can enhance their economic sustainability in the long run while minimizing their ecological footprint. One of the critical managerial lessons here is the value of financial investment and planning for CE.

Technology adoption is the priority of managers and organizations where integration of digital features, IOT, and artificial intelligence can offer transparent supply chain and better optimization of resources. However, organizations need to manage barriers related to technological infrastructure and collaboration with industries that are already good in this. Developing a technical landscape will help companies to move to circular process.

Further engaging in government policy and understanding the compliance can allow managers to view the opportunities for transparent CE process. Ensuring connecting sustainability and compliance with requirement of global environmental requirements can enhance corporate regulatory compliance. Also, the chapter discussed that managers should understand the importance of stakeholder collaboration. This will allow the successful implementation of CE. Organizations can adopt a well-thought collaboration with suppliers, NGOs, and government organizations. Also, they can work together with other sectors in partnerships for knowledge sharing and working as per the trend. This will allow businesses to frame creative strategies for enhancing their contribution to circularity.

9.4 Conclusion

This chapter offers significant inputs for CE and CS integration, adoption, and implementation. The chapter identifies some key challenges that hurdle the efficient process of these practices. These barriers are summarized as economic, technological, policy and regulatory, and stakeholder engagement barriers. The chapter discussed the essential elements of these challenges which are required to be addressed by the organizations. In manufacturing units, high initial costs are found to be the major financial burden. All manufacturing companies cannot afford the high feature technology which also limits their interest and practices for CE and CSR. Limited government incentives and tax relief are also demotivating factors for CE uptake. Regulatory uncertainty and weak enforcement discourage CE adoption. Disconnected policies that are nonspecific to the ideas of CE characterize the majority of Asian economies. However, CSR-based activities such as sustainability-linked funding and moral business activities enable organizations to attract responsible investors and create sustained economic value. The chapter highlights that it is important to address these issues which can encourage manufacturing companies to adopt CE and CSR efficiently.

References

Anbumozhi, V., & Kim, J. (2016). Towards a Circular Economy: Corporate Management and Policy Pathways. Research Papers in Economics. https://econpapers.repec.org/bookchap/eraeriabk/2014-rpr-44.htm

Bansal, P., & DesJardine, M. R. (2014). *Business sustainability: It is about time.* Strategic Organization, 12(1), 70-78. Sage Publications.

Bocken, N. M. P., de Pauw, I., Bakker, C., & van der Grinten, B. (2017). *Product design and business model strategies for a circular economy.* Journal of Industrial and Production Engineering, 33(5), 308-320. Wiley.

Chhimwal, M., Agrawal, S. and Kumar, G. (2022), "Challenges in the implementation of a circular economy in manufacturing industry", Journal of Modelling in Management, Vol. 17 No. 4, pp. 1049-1077. https://doi.org/10.1108/JM2-07-2020-0194

De los Rios, I. C., & Charnley, F. J. (2017). *Skills and capabilities for a sustainable and a circular economy: The changing role of design.* Journal of Cleaner Production, 160, 109-122. Elsevier.

Du, S., & Vieira, E. T. (2012). *Striving for legitimacy through corporate social responsibility: Insights from oil companies.* Journal of Business Ethics, 110(4), 413-427. Springer.

Ferreira, F. A., Jalali, M. S., & Ferreira, J. J. (2021). *Sustainability, a circular economy, and corporate social responsibility: Mapping the interconnections.* Sustainability, 13(2), 865. MDPI.

Fobbe, L., & Hilletofth, P. (2023). Moving toward a circular economy in manufacturing organizations: the role of circular stakeholder engagement practices. *The International Journal of Logistics Management*, *34*(3), 674-698.

Goyal, S., Rahman, Z., & Kazmi, A. A. (2023). *Corporate social responsibility and sustainable business practices: A strategic perspective.* Business Strategy and the Environment, 32(4), 1017-1033. Wiley.

Holly, F., Kolar, G., Berger, M., Fink, S., Ogonowski, P., & Schlund, S. (2023). Challenges on the way to a circular economy from the perspective of the Austrian manufacturing industry. Frontiers in Sustainability. https://doi.org/10.3389/frsus.2023.1243374

Holly, T., Smith, J., & Williams, R. (2023). *Financial barriers to circular economy implementation in the manufacturing sector.* International Journal of Sustainable Business, 20(3), 214-229. Emerald.

Jakhar, S. K. (2017). Stakeholder engagement and environmental practice adoption: The mediating role of process management practices. *Sustainable Development, 25*(1), 92-110.

Kumar, R., Singh, K. and Jain, S.K. (2020), "An empirical exploration of Agile manufacturing for enhanced business performance in Indian manufacturing industry", World Journal of Science, Technology and Sustainable Development, Vol. 17 No. 1, pp. 90-111. https://doi.org/10.1108/WJSTSD-07-2019-0040

Kumar, V., Rahman, Z., & Kazmi, A. A. (2016). Assessing the Influence of Stakeholders on Sustainability Marketing Strategy of Indian Companies. *SAGE Open, 6*(3). https://doi.org/10.1177/2158244016667991 (Original work published 2016)

Mohamed, K. (2024). *Long-term economic benefits of a circular economy adoption in manufacturing: A case study approach.* Journal of Business Sustainability, 18(1), 75-92. Emerald.

Nandha Gopan, S. and Balaji, M. (2023), "Indian automotive supply chains: barriers to a circular economy for sustainable development", Management Decision, Vol. 61 No. 11, pp. 3589-3609. https://doi.org/10.1108/MD-03-2023-0435

Rao, P., Kumar, S., & Patel, R. (2024). *Economic challenges and technological gaps in a circular economy adoption: A critical review.* Journal of Cleaner Production, 312, 127483. Elsevier.

Rao, Y., JP, R., Thomas, S., Agrawal, D., S .M, U. D., & Hajra, V. (2024). Circular Economy Strategies for Resource Efficiency and Sustainable Development in Manufacturing Industries. South Eastern European Journal of Public Health, 14–29. https://doi.org/10.70135/seejph.vi.2576

Sahoo, S., & Jakhar, S. K. (2024). Industry 4.0 deployment for a circular economy performance—Understanding the role of green procurement and remanufacturing activities. *Business Strategy and the Environment, 33*(2), 1144-1160.

Saidi, N. A., & Radzi, N. A. (2023). The Trend of a Circular Economy Studies in Asian Countries: A Bibliometric Analysis. International Journal of Academic Research in Economics and Management Sciences, 12(4). https://doi.org/10.6007/ijarems/v12-i4/20147

Salman, F., Brown, L., & White, M. (2024). *Regulatory complexities in the transition to a circular economy: A cross-sectoral analysis.* Environmental Policy and Governance, 34(2), 56-78. Wiley.

Salman, S., Taqi, H. Md. M., Nur, S., Awan, U., & Ali, S. M. (2024). The pathways to lean manufacturing for a circular economy: Implications for sustainable development goals. https://doi.org/10.1108/jrpc-10-2023-0009

Singh, R., Joshi, A., Raghuveer, K., & Kumar, V. (2025). Evaluating HRM practices for embracing the digital a circular economy in Indian manufacturing firms: an analytical study. *Strategy & Leadership*, *53*(1), 49-72.

Vasilev, V., & Nikolova-Alexsieva, V. (2024). Exploring the possibilities of integrating the a circular economy into sustainable development strategies through corporate social responsibility. *E3S Web of Conferences*, *585*, 11002. https://doi.org/10.1051/e3sconf/202458511002

Zaidi, A.A. and Chandra, R. (2024), "The challenges to a circular economy in the Indian apparel industry: a qualitative study", Research Journal of Textile and Apparel, Vol. ahead-of-print No. ahead-of-print. https://doi.org/10.1108/RJTA-09-2023-0105

10

Sustainable Urban Air Mobility: Challenges, Innovations, and Strategic Recommendations

Mamta Ahuja[1], Avani Bhatt[1], Priti Sharma[1], and Manu Sharma[2]

[1]Department of Commerce, Graphic Era Deemed To be University, India
[2]Department of Managememt Studies, Graphic Era Deemed To be University, India
E-mail: mamtaahuja20412@gmail.com; avanibhatt1188@gmail.com; pritisharma@geu.ac.in; manu.sharma@geu.ac.in

Abstract

Urban air mobility (UAM) is rapidly emerging as a promising solution to alleviate urban congestion, reduce transportation time, and enhances the sustainability of metropolitan environments and ensuring proper waste management practices for a healthy living environment. However, integrating UAM into existing urban infrastructures presents complex challenges across technological, regulatory, safety, environmental, societal dimensions, and waste management. This research aims to provide a comprehensive analysis of the key obstacles hindering the large-scale deployment of UAM, including technological limitations of electric vertical take-off and landing (eVTOL) aircraft, airspace congestion, regulatory uncertainties, and public acceptance. The study delves into the current state of eVTOL development, examining issues related to infrastructure, regulatory framework, safety standards, and system reliability. UAM involves aerial transportation systems such as drones and eVTOL aircrafts which can also in turn help cities in waste management in urban as well as rural areas. UAM is not just for passenger transportation with ease but also has its contribution toward public service operations such as waste collection and disaster response. But inculcating UAM for the use of waste management practices has its own challenges which India needs to

DOI: 10.1201/9788743806905-10

overcome. Through case studies and various references, the chapter identifies the potential challenges initiatives aimed at addressing these barriers, such as simulation-based air traffic management systems and impact of UAM in waste disposal methods. The research concludes by offering recommendations for overcoming these challenges, emphasizing the need of public-private partnerships and long-term sustainability in UAM development for providing ease in sustainable transportation and efficient waste management.

Keywords: Urban air mobility (UAM), eVTOL, regulatory challenges, waste management, sustainability, environmental impact.

10.1 Introduction

The rapid pace of urbanization has brought unprecedented challenges to modern cities, including traffic congestion, pollution, and inefficiencies in traditional ground transportation systems (Onokala and Olajide, 2020). As urban populations grow and demand for mobility increases, sustainable, innovative solutions, waste disposal management for a clean environment and public health are urgently needed. UAM has emerged as a promising frontier in addressing these challenges by introducing a three-dimensional transportation model that leverages electric vertical take-off and landing (eVTOL) aircraft, autonomous drones, and other advanced aerial technologies. Sustainable urban air mobility (SUAM) not only envisions a future where transportation is faster and more efficient but also prioritizes environmental stewardship, public safety, social acceptance, and sustainable waste solutions. By offering a greener alternative to conventional mobility systems and acting future potential tool and technology for dealing in major issues of waste management (Choudhary and Dahiya, 2023), SUAM aligns with global objectives, such as the United Nations Sustainable Development Goals, particularly those focusing on sustainable cities and communities. This technology has its impact majorly in providing sustainable transportation and waste management solutions. UAM can enhance waste management in several ways by deploying its technology in hard-to-reach areas to collect lightweight waste or hazardous materials. UAM vehicles can be frequently utilized for a variety of purposes, including serving as multifunctional assets in waste management and urban logistics, such as helping in quick transport of medical waste from hospitals or disaster-hit areas. These vehicles can be highly useful when it comes to post disaster cleanup by assisting in clearing debris and transporting essential supplies. Despite its transformative potential, the integration of SUAM into existing urban ecosystems is fraught with challenges.

UAM is a disruptive new sustainable air transportation system that is safe and secure. It enables efficient and reliable aerial transmission and ameliorates escalating traffic congestion in and around urban areas; along with this, there is huge scope of harnessing UAM technology in the fields of waste management. The integration of UAM in waste handling techniques is still far behind but its impact can be enormous if India overcomes the above-mentioned challenges. UAM can be widely used in small as well as large-scale waste collection. Electric vertical take-off and landing (eVTOL) aircrafts are seen as a long-anticipated enabler for UAM, providing a greener, quieter, and more sustainable mode of transportation (Xiang et al., 2024). It can also be used to make a significant impact in the waste disposal management. These aircrafts (eVTOL), often referred to as flying taxis, are designed to hover, take off, and land vertically while being powered by electric propulsion (motors, batteries, and fuel cells) systems (Jahn and Choueiri, 2003). UAM is a developing aviation concept that NASA characterizes as a "secure and efficient air transportation system that enables small package delivery drones and passenger-carrying air taxis to operate above defined areas" (NASA). A comprehensive examination of the historical development and operational details of the UAM concept is necessary. Air-based urban transport has been under development since the early 1940s, when researchers first investigated this concept. Between 1947 and 1971, Los Angeles Airways (LAA) employed helicopters to transport passengers and correspondence through a variety of locations in the Los Angeles Valley. However, the services were deferred in the aftermath of the two accidents attributable to mechanical malfunction in 1968 and subsequently ceasing the operations (Harrison, 2017, as referenced in Thipphavong et al., 2018, Garrison et al., 2021).

Technological advancements have significantly enhanced UAM. The present situation regarding UAM and its continuous advancement illustrates a swiftly transforming industry, with numerous nations formulating strategies for its implementation. In the United States, the NASA has developed a mission for advanced air mobility (AAM). UAM is a subset of the AAM idea being developed by the FA. UAM, a subclass of AAM, focuses on the transportation of passengers and freight inside urban metropolitan regions. NASA has announced the operational status of the electric vertical take-off and landing (eVTOL) urban flight network, facilitating the safe and serene operation of both unmanned and manned vehicles for on-demand and planned flights. In addition to conventional aircraft, the eVTOL concept aims to function with improved precision and consistency, facilitating a shift from two-dimensional transportation and logistics to three-dimensional mobility (Zhang et al., 2024).

Joby Aviation, Sarla Aviation in Bangalore, Lithium in Germany, Volocopter in Dubai, and Airbus in Brazil are among the numerous corporations worldwide that have dedicated recent years to the development and testing of eVTOLs that could facilitate air taxi operations.

Cutting-edge technologies in the near past have led to the advancement of UAM, an alternative and environment-friendly approach compared to conventional ground transportation systems (Cho and Kim, 2022). However, it is postulated to offer assorted opportunities to accomplish sustainable growth because of its novel transportation system concept, it fails to eradicate certain gaps faced within the Indian horizon. Despite different model definitions, UAM is caveated by many aspects mainly related to the five key pillars: technology, regulation, infrastructure, environmental impact, and public acceptance (EASA, pwc, 2023).

This research study provides insights into several constraints concerning the introduction and deployment of UAM to promote the sustainable cities and communities proposal established by the United Nations General Assembly_in 2015, "making cities inclusive, safe, resilient and sustainable" (UN). India is a country with a population of 1.45 billion, and the urban population is 47.6% of the total population. As the population increases, the traffic congestion directs the requirement for a more advanced transportation system. India generates around 62 million tons of waste each year, including both recyclable and nonrecyclable waste. More than half of the SDGs implicitly or explicitly address waste management. Hence, addressing its challenges and achieving our sustainable development goals has become a crucial aspect addressed through research and development efforts. This chapter studies an overview of the adoption and experimental phase of UAM globally but is primarily concerned with India's position at this juncture, highlighting the challenges that the Indian economy may face during the adoption and implementation of UAM. Furthermore, UAM sustainability and recommendations on the interventions required for UAM have been explored.

10.2 Literature Review

Since the emergence of the concept of UAM, there has been a surge in studies exploring its potential to transform urban transportation systems. UAM is seen as a game-changing development, promising to reduce congestion, enhance mobility, and contribute to the decarbonization of transportation, surveillance, monitoring, analysis of waste sites (Mager and Blass, 2022), and emergency waste transportation. As UAM technologies, such as electric vertical take-off and landing (eVTOL) aircraft, begin to mature, various frameworks have been employed to understand the potential impacts, particularly in terms of demand forecasting, infrastructure needs, and societal acceptance. UAM in India proposes a promising solution to urban congestion, pollution, waste management, and mobility challenges. The adoption of electric vertical take-off and landing (eVTOL) aircraft and drones for passenger and cargo transport is gaining traction worldwide. Transporting bulk of waste from collection centers to landfills, recycling UAM vehicles can reduce road congestion and improve waste transport efficiency in cities. India, with its rapidly urbanizing population and traffic problems, presents a unique opportunity and drawback simultaneously for UAM.

With the moving demand for sustainable mobility, there is a growing imperative need for UAM addressing a paramount shift toward environmentally responsible approaches and underscoring the need for innovation that reduces environmental impact. The number of passengers in India has doubled in the last 15 years, and this considerable increase in traffic is affecting the country's economy. According to the World Traffic Index Report (2024), these four major urban and metropolitan cities (New Delhi, Mumbai, Bangalore, and Hyderabad) have been among the 10 most choked roads in the world and are consistently listed on the list. UAM systems powered by electric take-off and landing (eVTOL) can provide alternative sustainable transportation options, thereby reducing traffic congestion. These systems can also be used as aerial waste transport, thereby reducing reliance on large garbage trucks, cutting down fuel consumptions and urban congestion. Air pollution is a major concern in India, with various Indian cities ranking among the most polluted areas in the world. UAM, which uses electric vertical take-off and landing (eVTOL), relies on greener technology like electricity and new decarbonized fuel, thereby reducing greenhouse gas emissions and fuel consumption, leading to a low carbon footprint (Cho and Kim, 2022).

10.2.1 UAM: Technological breakthroughs

UAM offers an innovative transportation solution in today's era, where prioritizing the environment has become essential, leveraging groundbreaking technology that is a safer, faster, cleaner, and more sustainable way to travel. UAM technology encompasses advanced aircraft and autonomous flight systems using real-time data, ensuring better navigation and better airspace management systems, aerial waste collection and transport. UAM accommodates a variety of "advanced technologies such as electric propulsion, high-tech materials, and sensors for navigation and safety" (Arafat and Pan, 2024). A technological breakthrough of UAM lies in the innovation of eVTOL (electric vertical take-off and landing) aircraft. These technologies are designed to operate more efficiently, quietly, and with little but minimal environmental impact than traditional aircraft. Although there is still some time away for UAM to become an everyday and comfortable mode of transportation for the public, it carries enormous potential.

10.2.2 UAM: Global scenario

Over the past few decades, urban pollution density has increased drastically, with unprecedented traffic congestion, commuter delays, and generation of excessive waste. "By 2050, approximately 80% of the European and North American populations are estimated to live in metropolitan areas causing major cities worldwide to grapple with increasing infrastructure and mobility challenges at the same time due to this there has been a serious impact on both human and environmental health" (Arafat and Pan, 2024). Therefore, to tackle such challenges, UAM comes out to be a transformative approach, an innovative and sustainable transportation aiming to provide short-distance passenger transport within cities or nearby cities. The adoption of this innovative solution globally is advancing smoothly. In North America, various companies such as Joby Aviation, Boeing, and Archer are developing these aircraft. Joby Aviation successfully conducted its first FAA testing under type inspection authorization (TIA), now in the final phase of the certification program (AAM International, 2025). "Joby Aviation completed a series of test flights in Korea as a part of K-UAM Grand Challenge" (Joby Aviation, 2024).

"Joby Aviation's first air taxi to become the first to conduct demonstration flights in Goheung, Korea, as a part of the country's K-UAM Grand Challenge aiming to advance the commercialization of air taxis in Korea."

Source: Joby Aviation

Saudi Arabia's ambition to reach the net zero emissions by 2030 through there "Saudi Green Initiative 2030" has been adopting "Waste Management solutions" by integration of UAM technology to detect the location, quantify waste through aerial images (Reddy et al., 2024). European countries are collaborating under programs like SESAR (Single European Sky ATM Research) to establish UAM standards. UAM's global adoption is a testament to its potential to revolutionize urban transportation. As countries invest in technology and infrastructure and most importantly, seek regulatory approval, UAM will play a crucial role in developing future smart cities worldwide.

10.2.3 UAM: In the contemporary Indian context

UAM comes out to be a largely untapped opportunity in the Indian landscape with a population of more than 1.45 billion, the urban population has increased almost six times since 1951 and it is estimated that 590 million people are expected to reside in the Indian cities by 2030, which is twice the population of USA. India's demography largely consists of a young population (under 35 years). As the population increases, so does the need for a more transformative and advanced transportation system. As per the Boston Consulting Group Study, urban congestion is expected to cost India almost USD 37 million by 2030 (TOI, 2024). The UAM Expo 2025 is expected to shape the future of UAM and is poised to "unite stakeholders from across the globe to revolutionize short-haul air mobility. It has been organized by the Confederation of Indian Industry (CII) partnering with Hunch Mobility is a part of Bharat Mobility Global Expo 2025, a leading platform for unveiling global mobility solutions this critical event is backed by the Government of India and the Ministry of Aviation is to be held from January 19-22, 2025, in Greater Noida (BLADE, 2025)".UAM in India is currently in its nascent stage, attracting both the public and private sectors.

10.2.4 UAM: A sustainable solution

UAM symbolizes a forward-looking approach to addressing unique urban mobility challenges while prioritizing sustainability. Through the integration of clear and faster technologies, the establishment of more advanced urban spaces, and, most importantly, aligning them with renewable energy sources, UAM is likely to become a cornerstone for future smart cities and communities, as well as reshaping the traditional urban transportation system into an eco-friendly and efficient ecosystem. One of the core advantages of UAM lies in its environmental impact. UAM's heavy reliance on eVTOL and drones powered by electricity rather than fossil fuel reduces carbon emissions and contributes to global decarbonization goals. Monitoring air quality in and around landfills has become essential today for aiming a cleaner and safer environment for living, with the help of eVTOL aircrafts which are equipped with sensors facilitating real-time data collection of pollutants such as methane and carbon dioxide (Surface Emission Monitoring). By using aerial pathways, UAM reduces traffic congestion, thereby saving people's time and money, and improves air quality in cities.

10.3 Main Text

10.3.1 Public acceptance

Customer acceptance and their propensity to pay for technology adoption are critical factors in the success of innovative products. The public accepts them when the benefits are readily apparent. Before UAM can become the next iteration of transportation technology, it must enhance customer acceptance. Numerous studies have investigated the factors that induce consumers to employ UAM. The adoption behavior of UAM consumers is contingent upon the quality of service and the level of trust, as per Kim et al. (2022).

The UAM technology is experiencing limited adoption due to its status as a recently developed system that requires proven solutions to address its public concerns before it can be extensively adopted. Customers will likely be concerned about the reliability and operation of eVTOL bikes, as well as the impact they have on infrastructure projects and the environment, as safety protocols will likely compel them to adopt them.

The CORUS-XUAM project began the process of developing potential mitigation strategies by reviewing existing drone acceptance surveys in order to gain a better comprehension of these public and environmental issues. The research aims to address public concerns about drone adoption by establishing prevention frameworks that encourage the acceptability of drones in urban airspace (Çetin et al., 2022). A technological roadmap necessitates proactive campaign organization and public education campaigns, as well as integrated public feedback, to ensure the correct development of the technology.

By analyzing the three groups of influencing factors as illustrated in Figure 10.1 – domain-specific, target-specific, and environment-specific – one can derive the concerns regarding accepting UAM. Domain-specific factors refer to the personal attributes or behavioral traits of individuals who might contemplate UAM as a mode of transportation. In contrast, target-specific factors consider an individual's assessment of technology (i.e., the object) within the respective environment. Environment-specific factors pertain to the conditions (subjects) that may influence an individual's acceptance of this technology (Babetto et al., 2023).

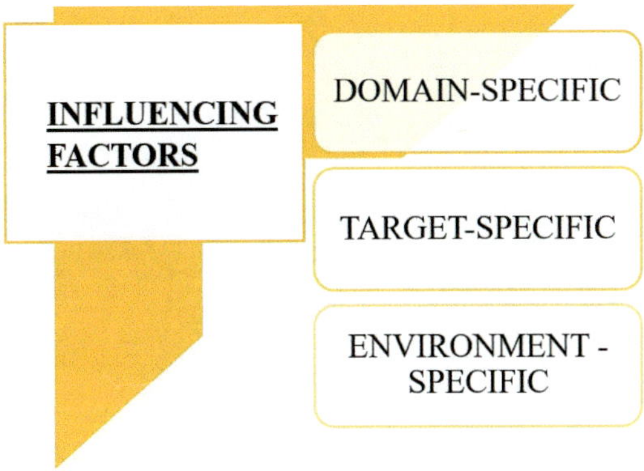

Figure 10.1 Groups of influencing factors for primary conventional acceptance.

Domain-specific factors include psychological, sociological, and demographic characteristics of individuals. It has been observed that age and gender are two very important factors influencing commercial air vehicle acceptance (Fu et al., 2019). Younger and male respondents are generally more receptive to this technology. The sociological factors that may comprise employment status, education, and culture can impact an individual's selection of mode of transportation. The object-related factors include safety, reliability, noise, privacy, and cybersecurity concerns (NASA, 2018). Environment-specific factors refer to wider social, economic, and cultural conditions in which UAM might be implemented (Babetto et al., 2023).

Figure 10.2 summarizes the overall factors that influence a person's choice of mode of transport and acceptance of UAM vehicle systems. Noise and safety are considered the most critical factors and are also leading factors by margin. Privacy and environmental impact are the factors that people nowadays give the utmost importance and value. Enhanced mobility choices refer to additional and more reliable sources of modes of transportation (EASA, 2021).

10.3.2 Safety and reliability

The most significant challenge to the successful incorporation of aerial vehicles into everyday urban transportation is the complex issue of UAM safety and dependability. In order to operate safely in hazardous dynamic environments, eVTOLs and other vehicles classified under UAM require fully operational advanced propulsion systems, autonomous flight controls, and battery-powered energy sources. The design mission established through

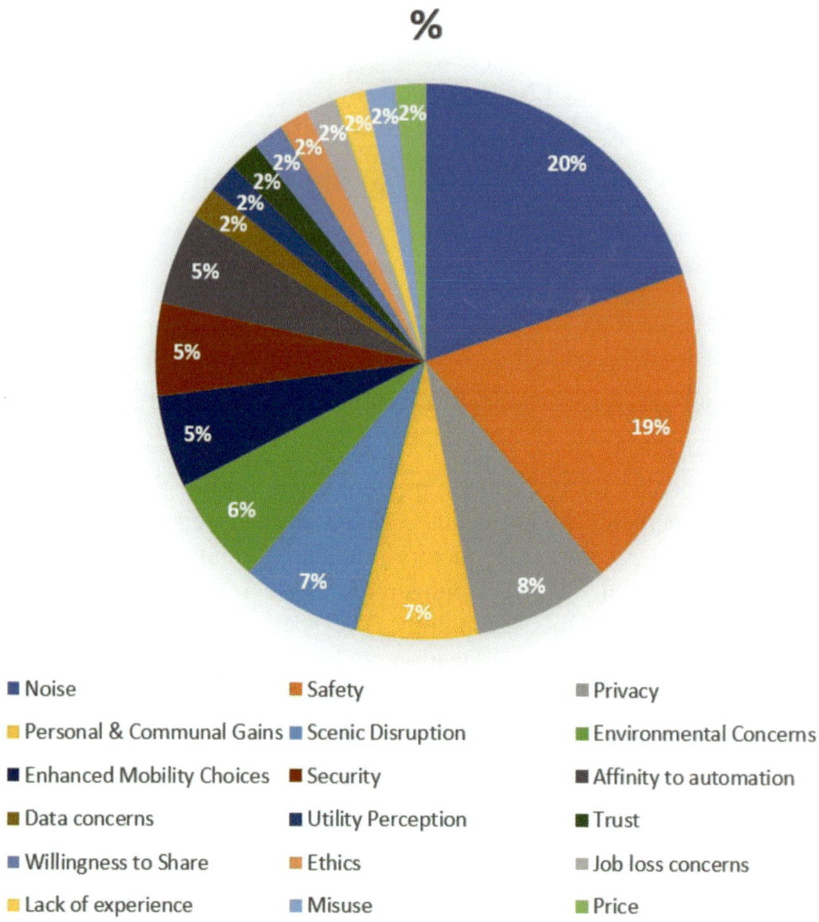

Figure 10.2 Public acceptance factors.
Source: EASA

NASA's NDARC (NASA Design and Analysis of Rotorcraft) (NASA) is adopted by the UAM safety assessment.

In terms of safety, ensuring that these vehicles can safely operate in crowded air spaces, avoiding other aircraft, buildings, power lines, and other urban infrastructure, is a major concern. Additionally, any malfunction – whether it involves propulsion or batteries – could have catastrophic consequences in urban areas where emergency landings are difficult, and densely populated areas accumulate risk (Mishra and Palanisany, 2023). Reliability is critical as UAM systems must perform consistently under real-world conditions, which include unpredictable weather, fluctuating power demands, and potential system failures. For instance, electric propulsion systems and batteries must be durable and able to operate over extended periods without significant degradation, while redundant systems must be in place to mitigate any failure of critical components.

10.3.2.1 Safety challenges

a) *Aircraft failures and redundancy:*
UAM vehicles must be designed with fail-safe systems and redundancy in critical components (for example, propulsion, power supply, and navigation systems) to minimize the risk of accidents. For example, if one of the electric motors in an eVTOL fails, the vehicle must still be able to land safely. Achieving this level of reliability requires advanced engineering, testing, and compliance with aviation safety standards and enhancing navigation integrity for UAM with redundant inertial sensors (Wang et al., 2022).

- Challenge – Ensuring that UAM aircraft have enough redundancy and fail-safes to operate safely even in the event of system failures is critical. Given the complexity of these systems, creating reliable, cost-effective redundancy is a technical challenge.

b) *Airspace integration and traffic management:*
UAM systems must integrate into existing airspace structures or operate within newly designed air corridors to avoid collision with other aircraft, including conventional aviation (for example, commercial aircraft and helicopters) and drones. As air traffic becomes more congested, especially in urban areas, ensuring safety through real-time coordination and conflict resolution will be essential (Xue, 2021).

- Challenge – Creating robust air traffic management systems for UAM that can handle the increased complexity of urban skies, ensuring safe separation between autonomous UAM vehicles and traditional air traffic (Becker et al., 2024).

c) *Autonomous flight and pilot intervention:*
Many UAM vehicles will be autonomous or semi-autonomous, reducing the need for human pilots. However, autonomous systems are not infallible and can be susceptible to software glitches,

censor malfunctions, or cybersecurity threats. Additionally, even with autonomous technology, human oversight may still be necessary, and pilot intervention in certain situations must be safe and effective.

- Challenge – Ensuring that autonomous UAM systems can respond to real-time hazards (for example, weather conditions, obstacles, and other aircraft) and implementing mechanisms for pilot interventions if needed.

d) *Passenger safety and emergency situations:*
For UAM to achieve mainstream adoption, passengers must have confidence that these vehicles can effectively handle emergency situations, including power failures, severe weather conditions, and accidents. This includes ensuring that the vehicles can make emergency landings safely and that passengers are adequately protected in the event of a crash (ANRA Technologies, 2022).

- Challenge – Designing emergency protocols, safety features (for example, parachutes, emergency landing systems), and evacuation processes that can be quickly and effectively executed in urban environments (West and Sherry, 2020).

10.3.2.2 Reliability challenges

e) *Battery and power systems:*
UAM vehicles are typically electric, relying on advanced battery systems to provide sufficient power for flight. However, current battery technologies face limitations in terms of energy density, weight, cost, and charging time. For UAM to be reliable, these vehicles must have enough power for long-range flight while maintaining a high level of productivity (Qiao et al., 2023).

- Challenge – Ensuring that battery systems are reliable and have a sufficient lifespan to handle frequent flights without the need for constant maintenance or replacements. Battery failure could lead to catastrophic consequences in flight, making reliability a major issue.

f) *System and software reliability:*
UAM vehicles rely heavily on complex software systems for navigation, communication, and flight control. Software glitches or malfunctions could lead to disastrous outcomes. Given that UAM systems may operate autonomously, a failure in the software could

result in serious accidents; thus, UAM software needs to make real-time decisions on route adjustments, collision avoidance, and emergency responses. Delays, errors, or inefficiencies in decision-making could jeopardize the safety of passengers and others on the ground. Software must be optimized for high-speed processing and decision-making (Justin et al., 2021).

- Ensuring the robustness of software, particularly in autonomous systems, so that it can respond safely and predicatively in the event of unexpected circumstances or malfunctions (Jin et al., 2024).

g) *Infrastructure and charging stations:*
UAM relies not only on the reliability of the aircraft itself but also on the auxiliary infrastructure, such as vertiports (take-off/landing zones), charging stations, and maintenance facilities. The infrastructure must handle frequent UAM vehicle use, which would lead to wear and tear, delays, or operational failures. The planning of vertiports profoundly influences the efficacy of the UAM system, since it represents the essential physical infrastructure that supports the entire operation of the UAM system (Brunelli et al., 2023; Rajendran and Srinivas, 2020).

- Developing a reliable, efficient, and widespread network of vertiports and charging stations that can support UAM operations at scale, minimizing downtime and ensuring that charging and maintenance are quick and efficient. Shin et al., 2022 introduced an optimization approach to ascertain the ideal sites for vertiports. The model accounted for traffic congestion to reduce travel time, the fixed cost of installing vertiports, operational costs, and collision risk expenses.

10.3.3 Investment and funding

UAM is poised to transform urban transportation in India by integrating electric vertical take-off and landing (eVTOL) aircraft, drones, and other air-based technologies into the country's transportation infrastructure. "Notably, Rajiv Bansal, Secretary of the Ministry of Civil Aviation, has boldly declared India's ambition to pioneer the eVTOL revolution, with a vision of realizing this global dream by 2025" (Tekriwal, 2024).

However, while UAM presents significant potential for reducing congestion, improving accessibility, and enhancing mobility efficiency, it faces substantial investment and funding challenges in India. Overcoming these challenges is vital for successfully deploying and commercializing UAM solutions.

h) *Uncertain return on investment (ROI):*
Investors are often wary of emerging technologies due to uncertainty around returns. While UAM has massive potential, it is still in the experimental and developmental stage, and the return on investment (ROI) is unclear, especially in a country like India with a unique socio-economic challenge.

i) *Lack of private sector investment:*
The private sector plays a crucial role in funding innovation and infrastructure, but many private investors in India are hesitant to take on the high risks associated with UAM. Private investment in aerospace in India has been limited compared to other sectors like the IT sector and manufacturing. Investors are likely to back technologies that offer immediate returns and have a proven track record, such as electric vehicles or renewable energy. UAM can expedite air transport and circumvent urban congestion through the advancement of new technology; however, these innovations must be collaboratively created alongside urban infrastructure (Rautray et al., 2022).

10.3.4 Infrastructure

The development and implementation of UAM in India holds a transformative potential for sustainable and efficient urban transportation, yet its advancement is met by complex infrastructural challenges. Quite a few companies globally have begun to concentrate on building the infrastructure necessary to service these aircraft. These comprise companies like VPorts, Aeroauto, Skyports, Volatus, Ferrovial, Varon, and Skyportz that design and operate vertiports. In the upcoming years, the UAM infrastructure market is predicted to grow significantly, with an estimated value of about $6.2 million by 2033. The successful deployment of UAM requires a well-built and extensive infrastructure. These include testing infrastructure, the number of vertiports, energy and charging/fueling stations, and communication infrastructure (KPMG, 2023). First and foremost is ensuring that a substantial number of test flights in a controlled environment is available to make UAM a reality. India requires defined airspace and testing zones where UAM systems can be rigorously tested without interfering with current aviation traffic. This might further require large areas of ground and air space for controlled trial of flight operations, communication systems, and safety protocols. Nevertheless, procurement of such might be quite capital-intensive for UAM technology developers, leading to legal impediments, and thereby requiring clearance and approval of regulatory authorities (pwc, 2023, Krylova, 2022). Additionally, as Indian cities continue to expand, integration of physical transport infrastructure specialized for take-off and landing locations known as "vertiports" must align with the existing urban fabric and consider how it can be redeployed with minimum physical modification (Cohen et al., 2021). Quite a few companies globally have begun to concentrate on building the infrastructure necessary to service these aircraft and are in the experimental phase. These comprise companies like VPorts, Aeroauto, Skyports, Volatus, Ferrovial, Varon, and Skyportz that design and operate vertiports. However, given India's current state, it remains in the early stages of progress. To manage and foresee levels of traffic, reliable access to energy and charging stations is essential. Establishing a dense network of high-capacity charging stations and refueling facilities (e.g., aviation fuel, hydrogen, and other alternative or biofuel) ensures that they are powered by renewable energy sources (Cohen et al., 2021). Finally, some other challenges include adapting and adopting communication systems. India's existing aviation infrastructure is not yet ready to manage the complexities and magnitude of UAM operations. UAM requires the implementation of advanced communication systems, navigation, and surveillance with well-built air traffic control (ATC) systems. It is essential to build such systems that can thoroughly support UAM communication requirements, including secure and interference-resistant links as well as mitigation mechanisms against jamming and spoofing (Safwat et al., 2024). UAM vehicles, including electric vertical take-off and landing (eVTOL) aircrafts, operate at a lower altitude, and robust ATCs are likely to ensure safe routes and separation of aircraft. The anticipated increase in the number of aircrafts will necessitate the provision of safe and convenient parking during peak hours within the range of areas of operations. Depending on the geographical locations, facilities may need to be covered with de-icing infrastructure (KPMG, 2023).

10.3.5 Environmental impact

UAM has been closely linked with electric-powered aircraft (eVTOL). Analyzing the lifecycle of UAM vehicles from production, operation, and disposal has a considerable effect on the environment. The manufacturing of these vehicles involves complex materials, some of which may not be recycled, thereby threatening the environment. Environmental concerns include noise and emissions. Although eVTOLs may not produce significant emissions during flight (Holden & Goel, 2016), the generation of electric power necessitates emissions of air pollutants, including greenhouse gases and other by-products from hydrocarbon combustion (Zhao et al., 2022). In the early years of operations, UAM will rely on helicopters and VTOLs powered by nonrenewable sources. Although quieter than helicopters, the noise from eVTOLs could still disturb nearby residential as well as commercial areas if UAM is used frequently (Cohen et al., 2024).

10.3.6 Regulatory framework

UAM in India presents a unique set of critical regulatory challenges that need to be addressed to pave the way for this emerging sector. A regulatory framework needs to be developed and adapted to ensure safety and efficiency. The deployment of eVTOL aircraft commercially requires three critical regulatory approvals that companies must obtain (see Figure 10.3) (Baker and McKenzie, 2022):

- Type certification: This is an initial step of the regulatory framework for the commercialization of UAM vehicles (eVTOL). Type certification involves obtaining regulatory approval for the operational fitness of a particular manufacturing design or category of their eVTOL aircraft for seeking this certificate.
- Product certification: Once the type certificate is pursued, the production certificate allows for the mass production of their respective design eVTOL model. Agencies or manufacturers producing such aircraft must

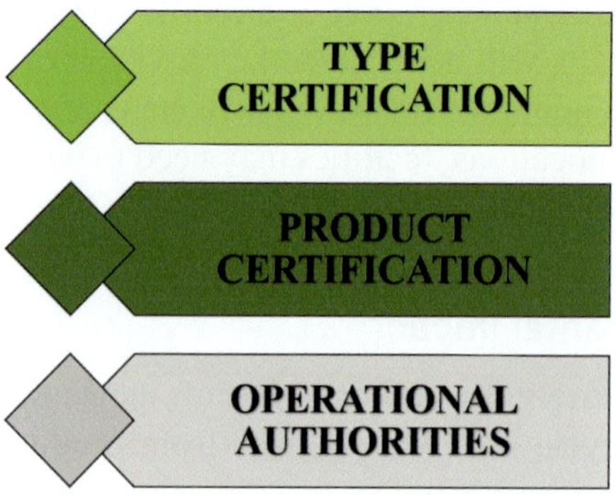

Figure 10.3 Three critical regulatory approvals.

ensure that all the guidelines and standards mentioned in the above type certificate are adhered to.

- Operational authorities: For the operation of UAM vehicles commercially, whether for producing passengers or cargo, additional operational requirements and authorization are necessary. These include safety, maintenance, performance, and other operational considerations.

To fully support UAM, India needs a robust regulatory framework that integrates safety protocols, infrastructure planning, noise control, environmental considerations, and privacy. One of the significant challenges in the integration of UAM into urban areas is noise pollution. Noise generated from increased electric take-off and landing (eVTOL) and other UAM vehicles has the potential to contribute to an increase in background acoustic levels in the urban environment, which may impact the quality of life of the neighborhood. Regulations must establish noise generation standards for acceptable UAM vehicles that are minimally disruptive to urban communities. Another factor that is of utmost importance to getting into the regulatory authority's prescribed standards is appropriate infrastructure. Regulations must set safety standards for infrastructure development and implementation of vertiports, including the locations, designs, passenger terminals, charging stations, and maintenance facilities. Coordination with the existing infrastructure is crucial in the initial phase as it requires various other challenges to be overcome. Cybersecurity and privacy pose another barrier. UAM vehicles might operate in overcrowded urban environments, leading to privacy concerns. The communities residing might feel uneasy and resistant to UAM if there are fears of being constantly monitored or recorded from above. Therefore, it is essential for authorities to explicitly tackle privacy concerns by creating definitive standards on how data is gathered and utilized by UAM operators. Environmental sustainability is necessary to implement and integrate eVTOL or UAM vehicles successfully. Regulatory bodies must ensure and establish emission standards considering not just the emissions produced during flights but also those generated from the manufacturing and disposal of nonrenewable waste from UAM vehicles and their infrastructure (Lundqvist et al., 2023).

10.4 Indian Scenario

The cities in India exhibit significant diversity regarding population, geographical areas, urban layout, topography, economic activities, income levels, and growth constraints. Consequently, the design of transportation systems must be thoroughly tailored to these specific characteristics. India is in

Table 10.1 Challenges of UAM technology and potential mitigation strategies.

Challenges	Threats	Probable mitigation strategy
Public acceptance	Noise generated by UAM vehicles, concerns about accidents, disruption to daily life	Design quieter UAM system vehicles using advanced propulsion and acoustic technology, communicate safety protocols clearly to the public, and use urban planning to reduce visual and operational disruptions
Safety and reliability	Personal safety, operational safety, cyber and physical threats, mid-air collisions, and system failure	Passenger evaluation system and background verifications, separate compartments, and emergency protocols. Execute thorough testing and certification of components and software
Investment and funding	High initial capital requirements, uncertain return on investment, regulatory delays, market uncertainty	Encourage public-private partnerships to share financial burdens, provide government grants, subsidies, and tax incentives for UAM development, detailed business, and revenue models, allocate contingency funds; conduct thorough market research
Infrastructure	Insufficient or poorly planned vertiports, high demand for charging stations, navigation, surveillance, information technology infrastructure	Develop scalable and modular vertiports to meet demand or integrate them with existing transportation (e.g., metro), upgrade grid capacity and storage of batteries in the ground to manage the power grid, and use eco-friendly building materials and sustainable construction practices
Environmental impacts	High noise levels, air quality degradation, and weather hazards pose safety risks and collisions with wildlife	Establish routes that bypass residential zones, formulate noise emission regulations, advocate for zero-emission planes, enhance safety measures and emergency protocols, create vehicles tailored for many weather conditions and climates, and deploy collision avoidance technology
Regulatory challenges	Lack of standardized regulations, slow regulatory adaptation, safety and certification delays, overlapping jurisdiction, lack of low-altitude airspace regulations, inadequate regulations addressing public concerns	Develop global standards in collaboration with international aviation (EASA, FAA), establish dedicated task forces to fast-track UAM-regulatory development, streamline certification processes for eVTOLs, include community input during regulatory development, and establish strict operational guidelines to address public concerns

the initial development phase in this emerging and disruptive alternative transportation sector (UAM). The European Union Aviation Safety Agency (EASA) and the Directorate General of Civil Aviation (DGCA) of India have signed a memorandum of understanding (MoU) regarding innovative air mobility and unmanned aircraft systems. This memorandum of understanding (MoU) establishes a joint civil aviation collaboration between the two aviation authorities regarding innovative air mobility systems and unmanned aircraft administration (MoCA, 2023). Numerous start-ups and aviation companies are investigating a possible collaboration with Indian authorities to introduce eVTOL in the country. For instance, Bengaluru International Airport (BIAL) and Sarla Aviation (SARLA, 2024) have partnered to set up India's first electric take-off and landing aircraft. Though India is gradually progressing in this sector and looking for a prospective alternative transportation solution, addressing its challenges will be crucial. India must adopt proactive measures and a strategic approach to effectively confront these challenges. India, home to almost 20% of the world's population, can present a huge market opportunity for UAM players. For the smooth adoption of UAM, India would require full-fledged eVTOL infrastructure, particularly vertiports, testing grounds, and charging stations to tackle potential risks associated with it. To mitigate such risks, the government should identify and develop designated areas for suitable establishment of testing facilities that will be utilized by private entities for testing eVTOL. Create a framework for assessing noise and environmental impacts that requires UAM companies to adhere to rigorous noise control regulations and emissions tracking, particularly in sensitive regions. This framework should incorporate policies from international frontrunners like Europe's EASA, which sets forth guidelines for noise and emissions concerning eVTOLs. Table 10.1 indicating challenges and proposed strategy for solutions.

10.5 Conclusion

UAM holds tremendous promise for alleviating India's urban congestion challenges to transform the future of transportation in densely populated cities requiring robust implementation of plans, holistic management strategies to face UAM threats which the country will face in the earlier as well as later years. UAM can be considered as a revolutionary prospect for India's transportation landscape, yet faces diverse challenges that require coordinated solutions across public acceptance, safety, reliability, investment, funding, infrastructure, regulatory, and environmental impact.

Infrastructure challenges have its own critical roadblocks which cause hindrance in UAM implementation. This research identifies a lack in the number of vertiports, testing infrastructure, air traffic control systems, and charging stations. The vertical integration of existing transportation modes also requires extensive work and planning that has yet to be addressed. The regulatory framework is something which requires a considerable amount of development, with significant gaps identified in areas of standardized regulations, safety standards, and certification delays highlighting, an urgent need for accelerated regulatory development. This research also highlights the need for further advancement to ensure safety, reliability, and investing and funding requirements. UAM's impact on the environment has also been discussed, including how it could pose serious threats to our planet, also highlighting how UAM can be considered beneficial in waste disposal management but requires much more advanced development. Perhaps the most crucial and challenging are the socio-economic factors of public acceptance issues.

Future authorities and policymakers must address all challenges with precision, balancing safety, accessibility, and aiming for a cleaner and healthier living environment. Collaborative efforts of the government, public-private partnerships, and research institutions are required for the successful deployment of UAM in India. UAM can play an essential role in revolutionizing urban mobility in India while contributing to the sustainable development goals.

References

Advanced Air Mobility (AAM) – NASA's active mission: Advanced Air Mobility Mission - NASA

Al Haddad, C., Chaniotakis, E., Straubinger, A., Plötner, K., & Antoniou, C. (2020). Factors affecting the adoption and use of urban air mobility. *Transportation research part A: policy and practice*, *132*, 696-712.

Anra Technologies - Safety of UAM – ANRA Technologies – Airspace and Mission Management

Babetto, L., Kirste, A., Deng, J., Husemann, M., & Stumpf, E. (2023). Adoption of the Urban Air Mobility System: Analysis of technical, legal and social aspects from a European perspective. *Journal of the Air Transport Research Society*, *1*(1), 152-174.

Becker, D., Schalk, L.M. Toward robust and efficient communications for urban air mobility. *CEAS Aeronaut J* (2024). https://doi.org/10.1007/s13272-024-00738-6.

Beyond roads: India embraces advanced air mobility technology for next-gen mobility in a $1.5 Trillion market - Express Computer

Brunelli, M., Ditta, C. C., & Postorino, M. N. (2023). New infrastructures for Urban Air Mobility systems: A systematic review on vertiport location and capacity. *Journal of Air T Transport Management*, *112*, 102460.

Çetin, E., Cano, A., Deransy, R., Tres, S., & Barrado, C. (2022). Implementing mitigations for improving societal acceptance of urban air mobility. *Drones*, *6*(2), 28.

Choudhary, R., & Dahiya, S. (2021, April). Drone Technology in Waste Management: A Review. In International Conference on Unmanned Aerial System in Geomatics (pp. 157-172). Cham: Springer International Publishing.

Cohen, A. P., Shaheen, S. A., & Farrar, E. M. (2021). Urban air mobility: History, ecosystem, market potential, and challenges. *IEEE Transactions on Intelligent Transportation Systems*, *22*(9), 6074-6087.

Cohen, A., & Shaheen, S. (2024). Advanced air mobility: Opportunities, challenges, and research needs for the state of California (2023-2030).

Democratizing Urban Air Mobility - Sarla Aviation

The Directorate General of Civil Aviation (DGCA) India has signed a Memorandum of Understanding (MoU) with the European Union Aviation Safety Agency (EASA) - Press Release: Press Information Bureau

Eissfeldt, H. (2020). Sustainable urban air mobility supported with participatory noise sensing. *Sustainability*, *12*(8), 3320.

EASA, 2021- Study of social acceptance of UAM in Europe - uam-full-report.pdf

Federal Aviation Administration - Urban Air Mobility (UAM) Concept of Operations 2.0_1.pdf

Fu, M., Rothfeld, R., & Antoniou, C. (2019). Exploring preferences for transportation modes in an urban air mobility environment: Munich case study. *Transportation Research Record*, *2673*(10), 427-442.

Garrow, L. A., German, B. J., & Leonard, C. E. (2021). Urban air mobility: A comprehensive review and comparative analysis with autonomous and electric ground transportation for informing future research. *Transportation Research Part C: Emerging Technologies*, *132*, 103377.

Hasan, S. (2019). *Urban air mobility (UAM) market study* (No. HQ-E-DAA-TN70296).

Harrison, 2017 - From the Archives: Los Angeles Airways helicopter overturns - Los Angeles Times

Holden, J., & Goel, N. (2016). Fast-forwarding to a future of on-demand urban air transportation. *San Francisco, CA, 2016.*

India Energy Storage Alliance (IESA) - https://indiaesa.info/resources/ev-101/4146-urban-air-mobilityJahnandChoueiri,2003-ElectricPropulsion-ScienceDirect

Jin, Z., Ng, K. K., & Zhang, C. (2024). Robust optimisation for vertiport location problem considering travel mode choice behaviour in urban air mobility systems. *Journal of the Air Transport Research Society*, *2*, 100006.

Justin, C., Patel, S., Bouchard, E. D., Gladin, J., Verberne, J., Li, E., ... & Kufeld, R. M. (2021). *Reliability and safety assessment of urban air mobility concept vehicles* (No. GTRI Document No. D9015A001R2).

Kim, M., & Cho, G. H. (2022). Examining the causal relationship between bike-share and public transit in response to the COVID-19 pandemic. *Cities*, *131*, 104024.

Kim, Young & Lim, Cherin & Ji, Yong Gu. (2022). Exploring the User Acceptance of Urban Air Mobility: Extending the Technology Acceptance Model with Trust and Service Quality Factors. International Journal of Human-Computer Interaction. 39. 1-12. 10.1080/10447318.2022.2087662.

KPMG, 2024 - Integrating air mobility into wider infrastructure - KPMG Ireland

Krylova, M. (2022). Urban planning requirements for the new air mobility (UAM) infrastructure integration. *Frankfurt University of Applied Sciences.*

Mager, A., & Blass, V. (2022). From illegal waste dumps to beneficial resources using drone technology and advanced data analysis tools: A feasibility study. *Remote Sensing*, *14*(16), 3923.

Miller, J., Jones, M., Klyde, D. H., Heffley, R., & Arencibia, A. (2022). Using Pilot Models to Support Urban Air Mobility Design, Analysis and Certification. In *AIAA AVIATION 2022 Forum* (p. 3942).

Mishra, S., & Palanisamy, P. (2023). Autonomous advanced aerial mobility—An end-to-end autonomy framework for UAVs and beyond. *IEEE Access*, *11*, 136318-136349.

Müür, J. (2024). CITYAM empowers responsible urban air mobility.

Onokala, P. C., & Olajide, C. J. (2020). Problems and challenges facing the Nigerian transportation system which affect their contribution to the economic development of the country in the 21st century. *Transportation Research Procedia*, *48*, 2945-2962.

PARK, S. W. (2021). *Social acceptability of urban air mobility by aircraft category and autonomous phases* (Doctoral dissertation, KDI School).

PWC, 2023- Advanced and short-haul Air Mobility - Microsoft Word - Draft Updated 4.docx

Qiao, X., Chen, G., Lin, W., & Zhou, J. (2023). The Impact of Battery Performance on Urban Air Mobility Operations. *Aerospace*, *10*(7), 631.

QED Environmental Solutions – Surface Emissions Monitoring

Rajendran, S., & Srinivas, S. (2020). Air taxi service for urban mobility: A critical review of recent developments, future challenges, and opportunities. *Transportation research part E: logistics and transportation review*, *143*, 102090.

Rautray, P., Mathew, D. J., Eisenbart, B., & Kuys, J. (2022). Understanding Working Scenarios of Urban Air Mobility. *Proceedings of the Design Society*, *2*, 563-572.

Reddy, B. H. V., Ramakrishnan, V., Almuaikel, M., Alanazi, K., & Alshaikh, A. A. (2024, November). Artificial Intelligence Associated Drones Solutions for Waste Disposal Management in the Process Industries. In *Abu Dhabi International Petroleum Exhibition and Conference* (p. D021S043R003). SPE.

Rice, S., Winter, S. R., Crouse, S., & Ruskin, K. J. (2022). Vertiport and air taxi features valued by consumers in the United States and India. *Case Studies on Transport Policy*, *10*(1), 500-506.

Safwat, N., Sabatini, R., Gardi, A., Hafez, I. M., & Newagy, F. (2024). Urban Air Mobility Communication Performance Considering Co-Channel Interference. *IEEE Transactions on Aerospace and Electronic Systems*.

The Sesar Project - The SESAR Project - European Commission

Shin, H., Lee, T., & Lee, H. R. (2022). Skyport location problem for urban air mobility system. *Computers & Operations Research*, *138*, 105611.

Thipphavong, D. P., Apaza, R., Barmore, B., Battiste, V., Burian, B., Dao, Q., ... & Verma, S. A. (2018). Urban air mobility airspace integration concepts and considerations. In *2018 aviation technology, integration, and operations conference* (p. 3676).

Times of India, 2024 - The future of mobility: Technological advancements solving urban congestion issues - Times of India

Urban Air Mobility - Urban Air Mobility (AAM) Simulation - NASA

UAM & Conditions for UAM - What is UAM | EASA

United Nations - Goal 11: Sustainable cities and communities - The Global Goals

Wang, S., Zhan, X., Zhai, Y., Zheng, L., & Liu, B. (2022). Enhancing navigation integrity for Urban Air Mobility with redundant inertial sensors. *Aerospace Science and Technology*, *126*, 107631.

West, J., & Sherry, L. (2020, September). Agent-based simulation of metropolitan area evacuation by unmanned air mobility. In *2020 Integrated Communications Navigation and Surveillance Conference (ICNS)* (pp. 6B1-1). IEEE.

Xiang, S., Xie, A., Ye, M., Yan, X., Han, X., Niu, H., ... & Huang, H. (2023). Autonomous eVTOL: A summary of researches and challenges. *Green Energy and Intelligent Transportation*, 100140.

Xue, M. (2020). Urban air mobility conflict resolution: Centralized or decentralized. In *Aiaa aviation 2020 forum* (p. 3192).

Yao, E., Guo, D., Liu, S., & Zhang, J. (2024). The role of technology belief, perceived risk, and initial trust in users' acceptance of urban air mobility: An empirical case in China. *Multimodal Transportation*, *3*(4), 100169.

Yedavalli, P., & Mooberry, J. (2019). An assessment of public perception of urban air mobility (UAM). *Airbus UTM: Defining Future Skies*, 2046738072-1580045281.

Zhang, J., Liu, Y., & Zheng, Y. (2024). Overall eVTOL aircraft is design for urban air mobility. *Green Energy and Intelligent Transportation*, *3*(2), 100150.

Zhao, P., Post, J., Wu, Z., Du, W., & Zhang, Y. (2022). Environmental impact analysis of on-demand urban air mobility: A case study of the Tampa Bay Area. *Transportation Research Part D: Transport and Environment*, *110*, 103438.

11

Sustainable Supply Chain Management through Smart Technology

Madhu Arora[1], Rakesh Kumar[2], and Yugank Arora[3]

[1]Symbiosis Institute of Computer Studies and Research, Symbiosis International (Deemed University), India
[2]Department of Management Studies, Graphic Era Deemed to be University, India
[3]Blue Heavens, Australia
E-mail: shamayutu@gmail.com; aigrakeshkumar@gmail.com; yugnakarora@gmail.com

Abstract

Sustainable supply chain management (SSCM) today has turned out to be mandatorily important for businesses that are trying to combine economic, environmental, and social performance. The sustainability and efficiency of supply chain are significantly enhanced by inclusion of smart technologies like the artificial intelligence (AI), the Internet of Things (IoT), blockchain, and big data analytics. Through the automation of decision-making, real-time tracking and updating, coupled with predictive analytics, smart technology enhances productivity of resources, reduces waste, and boosts transparency. AI-powered predictive models support reduced inventory costs, avert over-production, and estimate demand fluctuations accurately resulting in a positive environmental impact. Through traceability and end-to-end visibility owing to the utilization of IOT devices, real-time tracking of energy use during storage and transportation becomes a reality. Blockchain technology contributes in ensuring data integrity and traceability, thereby increasing stakeholder accountability at one end and confidence at the other. Moreover, through automation of ethical sourcing and meeting necessary compliances, smart contracts endorse social responsibility. Big data analytics increases

DOI: 10.1201/9788743806905-11

effective decision-making and continuous improvement through increase in performance of the supply chain. Thus, by these technologies incorporated in the business processes, firms will be able to create resilient and compliant supply chains that encourage achievement of sustainable development goals (SDGs). By presenting successful case studies and providing practical guidance to businesses wishing to implement sustainable practices, this study explores how smart technology is transforming SSCM. According to the findings, smart technology plays a major role in facilitating SSCM, encouraging corporate social responsibility, and producing both financial and environmental benefits.

Keywords: Supply chain management, sustainability, smart technologies, AI, IoT, blockchain.

11.1 Introduction

Considering the fast-paced global economy today, businesses have realized the importance of striking an equilibrium between economic progress and responsibility toward both the society and the environment. The flow and transformation of goods beginning at the extraction phase of raw materials to the consumer, as well as the course of communication involved, all encompass the supply chain. Information and supplies pass across the supply chain. With the aim of building a sustained competitive advantage, SCM assimilates these operations through boosted SC linkages (Seuring, 2008). Globalization puts pressure on supply chain management to address issues like fair labor practices and environmentally sustainable production that go beyond simple economic concerns. Concern over sustainable supply chain management is growing among businesses across different sizes and in diverse sectors. Reaching the minimal sustainability performance is ensured by adhering to social and environmental norms at every stage of the supply chain. The conception and unveiling of sustainable products can support the strategy of reacting to outside strain from governments, consumers, media, and the NGOs (non-governmental organizations).

According to Pagell (2014), SSCM has been proposed as a way to syndicate stakeholder apprehensions about profit with those about how a central firm's activities, both internal as well as supply management, affect ecological and social systems. Achieving environmental, social, or economic performance together with implementing ecologically and socially responsible behaviors are all considered sustainability outcomes. Adopting

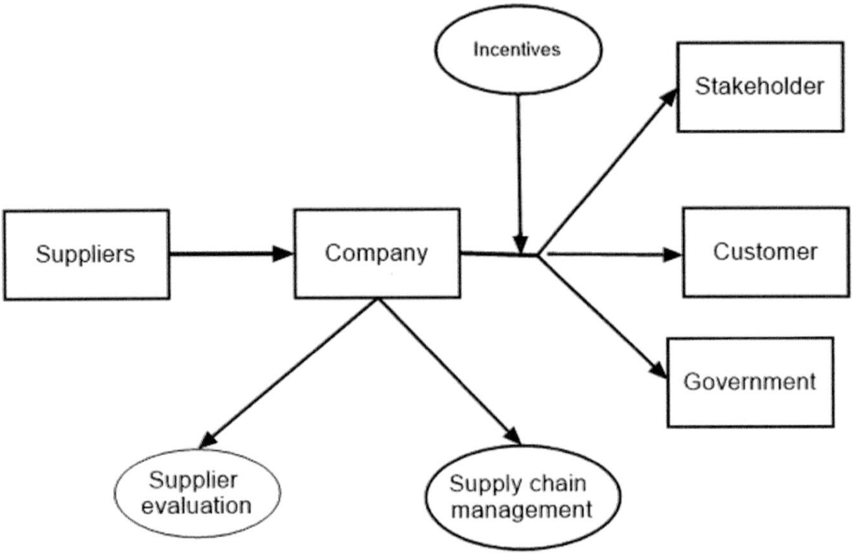

Figure 11.1 Prompts for SSCM (Source: adapted from Seuring, 2008).

environmental management systems, investing in pollution prevention and control, and obtaining environmental certifications like ISO14001 are examples of environmental practices. Adoption of social standards like ISO26000 and adherence to regional labor laws are examples of socially responsible actions. Environmental performance takes into account how well resources are used, and recycling, as well as effluence, waste, and emanations are reduced (Rao, 2005). Human privileges, labor practices, and the influence upon resident communities are all taken into account by social performance (Yawar, 2015). Sound, operative, or accounting-based metrics are employed to operationalize fiscal performance (Golicic, 2013).

Sustainable supply chain management, with its goal to minimize waste, promote ethical sourcing, and mitigate the environmental impact while retaining and enhancing productivity and profitability, helps meet the balance between growth and responsibility. The traditional supply chain (SC) processes have been transformed by the inclusion of smart technologies encompassing AI (artificial intelligence), IoT (Internet of Things), blockchain, and big data resulting in increased resilience and sustainability. Prompt for SSCM shown in Figure 11.1.

Digital products and services, together with management of SC operations in businesses, support use of smart technology in supply chains

(Büyüközkan, 2018). Increased trust, agility, and productivity are the results of the technological supply chain's increased information availability and vastly better interactions, communication, and collaboration. Smart technologies are entities that combine the intelligent features of digital technologies with physical equipment or processes. Programmability, addressability, sensibility, communicability, memorizability, and associability are some of the essential traits that constitute smart technologies and enable the interconnectedness and intelligence of businesses (Nasiri, 2020). Administrations need to give precedence to sustainable advancement and practice innovations including IoT and big data to form smart and robust SCs for sustainable functioning (Yousefi, 2022).

Application of smart technologies helps businesses optimize logistics, lower carbon footprints, and improve SC transparency through automation, real-time monitoring, and predictive analytics. Additionally, businesses can obtain a competitive edge with the help of cost reductions, increased productivity, and regulatory compliance, all made possible with the help of smart technology. Such technologies' advantageous consequences on assimilating varied supply chain facets, including suppliers, internal procedures, and customer relations, augment both the business's and the SC's performance (Verma, 2024). Using Industry 4.0 (I4.0) technologies makes it easier to implement circular development and make accurate decisions about supply chain operations, which in turn promotes sustainable supply chain management (SSCM) (Liu, 2023).

There are numerous examples that demonstrate how crucial smart technology is to the sustainable supply chain. The business's competitiveness, innovative performance, and operational efficiency have all increased as a result of the inclusion of smart technologies. Every activity in the tech-enabled SC, centred on Haier Group, demonstrates the critical role that smart technology plays in the SSC (www.sohu.com). World's initial "smart + 5G" platform COSMOPlat that supported mass customization, cutting-edge integration test, and verification was constructed by the Haier Group in 2019 (Pasi, 2020). The 5G technology in 2020's software version 4.0 decreased planning costs of fresh schemes, shortened the building cycle, and enhanced the equipment's IoT efficiency. Haier Group concentrates on using IoT, big data, blockchain, and 5G to enhance production measures while raising the caliber of its products.

Compared to traditional operations, smart technology-based operations are better able to recognizes various types of information in the surrounding environment and offer prompt feedback, updates, and improvements.

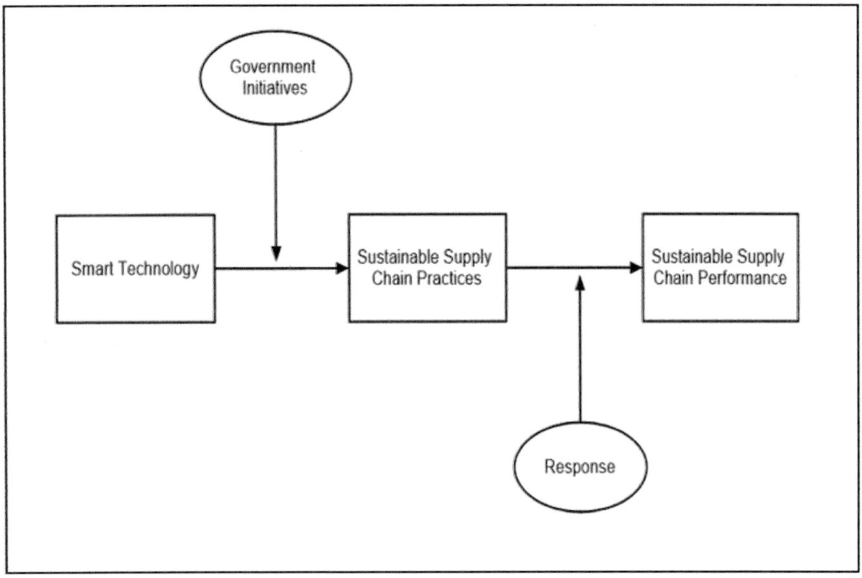

Figure 11.2 Smart technology and SSC correlation framework (adapted from Liu, 2022).

This chapter examines how supply chain management (SCM) and smart technology intersect, emphasising the advantages, difficulties, and prospects for firms looking to establish environmentally responsible and effective supply networks. Figure 11.2 shows smart technology and SSC correlation.

11.2 Sustainable Supply Chain Management through Smart Technology

11.2.1 Role of smart technology in sustainable SCM

In a highly dynamic and globalized world where continuous increase in production, elimination of waste while reducing negative influences on the environment has become an imperative, smart technology is certainly critical to SSCM. With the goal of establishing a link between digital alteration and business functioning, smart technologies are crucial (AlMulhim, 2021) as they act as an intermediary in the association between digital transformation and effective performance (Lee, 2024).

Data analytics, blockchain, IoT, and AI are some emerging technologies today that have immense potential to help businesses not only optimize their supply chains but also enhance sustainability to a great extent. Adopting such advances can help businesses save money, comply with regulations, as well as lessen their impact on the environment, ultimately resulting in strengthened and sustained supply networks.

11.2.2 AI & IoT integration for green SCM

AI and IoT are revolutionary in implementing green SCM strategy. To precisely forecast demand, systems equipped with AI can help examine market patterns, historical data, and current customer behavior to help lower surplus inventory and overproduction; reduce expenses incurred in storage and resource consumption. Lack of discernibility in the digital SC network made it tough for businesses with limited integration to improve sustainable performance (Saryatmo, 2021).

With the application of AI, transportation routes are optimized, which lowers both fuel usage and emissions. It does this by determining the best route for transit and identifying any blockages and suggesting alternative routes as well in that scenario. Also, automation powered by AI reduces waste by effectively monitoring and managing stock levels and avoiding overstocking or understocking. Organizations can also evaluate the sustainability of their suppliers by assessing carbon footprints of suppliers and ensuring adherence to the environmental regulations.

11.2.3 Blockchain for transparent and sustainable SCs

Blockchain is a distributed digital ledger that logs information safely and permanently. Each entry in a "block" is joined to the previous one to form a "chain" of data. Blockchains are therefore made up of both a distributed database and a data structure, which is a linked list of blocks that contain transactions. It is impossible to alter an earlier block without beginning the chain over from that point since each block is cryptographically chained to the one before it by containing its hash value and a cryptographic signature. Blockchain technology is linked to the concept of smart contracts, which are scripts that may read and write data from the blockchain and execute in response to particular transactions (Prause, 2019). This system is maintained across a network of computers (nodes) and ensures data transparency, consistency, and resistance to tampering.

Some key features of blockchain include: (i) decentralization with the data being shared across the network with no single entity have control over the data; (ii) immutability as a record cannot be modified or purged;

(iii) transparency with all participant having access to same data thereby increasing trust; and (iv) smart contracts which are self-managed contracts that automatically enforce terms to ensure predefined conditions being met.

Each stakeholder involved in the supply chain has access to single source of information. It allows for tracking each leg in the chain right up to delivery. Blockchain also allows for recording and tracking of each stage. This ensures real-time tracking of product origin. Businesses and consumers can use blockchain technology to confirm a product's origin. For instance, by keeping certification information on the blockchain, a coffee company may demonstrate that its beans came from fair-trade farms. By offering an unchangeable record of transactions, blockchain aids in the prevention of fraud. Products that are counterfeit can be promptly detected and eliminated from the supply chain. Manufacturers, suppliers, and distributors may work together more efficiently, cutting down on delays and misunderstandings, because all data is exchanged and updated in real-time.

Blockchain can monitor whether items come from sustainable and ethical suppliers. For instance, textile businesses can confirm that cotton is farmed without harming the environment and under fair labor circumstances. Blockchain can track how supply chain operations affect the environment, including trash production, energy use, and emissions. Businesses can utilize this information to accomplish sustainability goals and lessen their carbon footprint. Blockchain records the whole supply chain process, allowing for automatic adherence to labor and environmental laws. To make sure businesses adhere to industry norms, regulatory agencies might audit the data.

Blockchain promotes reuse and recycling by giving insight into product lifecycles. Businesses can monitor material returns and reuse, promoting the a circular economy paradigm.

11.2.4 Digital twins for supply chain optimization and sustainability

Digital twin is one cutting-edge technology that has enormous potential in solving various issues. A virtual product, procedure, or system that replicates its real-life matching part in the physical realm is called digital twin – a digital model of a real-world object. When employed in a supply chain, it becomes possible to achieve immense precision and agility in simulating, forecasting, and optimizing operations. In order to reflect a virtual model accurately depicting the state and functionality of real system, a digital twin gathers data from sensors, IoT devices, and other sources. It constantly keeps updating itself thereby allowing firms to: monitor performance, detect issues, predict future outcomes, and test scenarios without disrupting operations and optimized processes.

From gathering of raw supplies to distribution of finished goods, digital twins offer a thoroughly updated SC view. This helps businesses in monitoring inventory levels, deliveries, and supplier performance, thus resulting in increased visibility. Disruptions can be recognized well in time leading to a proactive response. Delivery times are reduced with streamlined logistics. "What-if" scenarios can be effectively examined for making better choices. This may include varying stock levels, modifying contracts with suppliers, adjusting to disruption, reducing risks and expenses by proactive action, reducing downtime by optimizing manufacturing schedules, and optimizing warehouse designs to boost productivity.

Digital twins can assist in estimation of demand with the help of machine learning (ML) algorithms that consider past data, market trends, etc. This enables firms in cutting down on overstocking and stockouts, increase client satisfaction, and ultimately reduce overproduction and thereby waste. By offering a consolidated perspective of the SC, digital twins help improve coordination and communication with partners and suppliers both. This helps in lead time reduction, accuracy in orders, and better relationship with suppliers. Digital twin therefore contributes toward reduction in waste and emissions; improved a circular economy practices by tracking lifecycle of products and material; enhanced energy efficiency; and finally supports compliances and regulations by providing exhaustive insights on environmental impact.

Businesses are increasingly under pressure to lessen their influence on the environment, encourage moral behavior, and improve long-term resilience, which has made sustainable procurement a top focus. While cost and efficiency were the main concerns of traditional procurement models, contemporary methods place more emphasis on the triple bottom line – balancing social, environmental, and economic performance.

11.2.5 Sustainable procurement

Sustainable procurement has become a priority as a result of growing demand on businesses to reduce their environmental impact, promote ethical behavior, and increase long-term resilience. Traditional procurement techniques prioritized cost and efficiency, but modern approaches focus more on the triple bottom line – balancing economic, social, and environmental performance. Smart technology allows organizations to assess environmental effects, monitor supplier practices, and automate decision-making to reduce waste, carbon emissions, and unethical labor practices.

By integrating these technologies into their procurement strategies, companies can reduce expenses, improve brand awareness, and adhere to legal requirements. Sustainable procurement is the process of making purchases of goods and services in a way that minimizes the impact on the environment, promotes ethical sourcing and fair employment standards, and ultimately supports long-term economic sustainability. It comprises choosing suppliers and products that promote social responsibility and environmental goals, as well as evaluating a product or service's entire lifecycle, from the extraction of raw materials to disposal.

Large datasets can be analyzed by AI-driven procurement tools to find trends, forecast results, and recommend the best sourcing tactics. Sustainable vendors are recognized through certifications, compliance records, etc. Hazards in the form of SC interruptions/environmental influence can be forecasted. Thus, sustainable procurement through use of smart technologies has numerous benefits including cost reduction, improved supplier relationships, enhanced brand reputation, reduced environmental impact, and improved regulatory compliance.

11.3 Environmental Impact and Sustainability

As global issues like pollution, resource depletion, and climate change continue to worsen, corporations are increasingly focusing on environmental impact and sustainability. In many industries, supply chains account for more than 90% of a company's carbon footprint, making them a significant contribution to environmental deterioration. Reducing supply networks' environmental impact is now a strategic requirement for long-term success. The smart technologies are today revolutionizing SC processes and leading to significant cut in waste, emissions, while transitioning toward a circular and sustainable economy. This section observes the practices through which sustainability can be upgraded and environmental impact drastically reduced owing to the usage of green warehousing, a circular economy models, sustainable packaging, and smart logistics.

11.3.1 Smart logistics

Logistics – comprising transportation, warehousing, and distribution – contributes significantly to carbon emissions. The logistics sector is responsible for nearly **14% of global greenhouse gas (GHG) emissions**. Smart logistics leverages technology to optimize transportation, reduce energy consumption, and minimize waste. Delivery routes are optimized using AI-powered logistics technologies that examine weather, traffic patterns, and fuel usage. Emissions and fuel consumption are decreased by shortening delivery routes and increasing load efficiency. Carbon emissions are reduced via the use of electric and hybrid delivery vehicles.By pooling transportation resources, companies are lowering the number of vehicles on the road through collaborative logistics and shared transportation.Fuel waste and empty miles are decreased by consolidated shipments. Warehouse-to-consumer direct shipping DTC models eliminate the need for middle-tier handling and warehousing. Reduced packing waste and energy consumption are the results of fewer touchpoints.

11.3.2 Smart warehousing

Smart warehousing is an infrastructure with low-energy use utilizing renewable energy sources, such as solar panels, to power storage facilities. Energy usage is decreased by installing smart HVAC (heating, ventilation, and air conditioning) systems and LED lighting. AI and machine learning make dynamic adjustments to inventory levels and forecast demand trends. Automated inventory control minimizes waste and emissions by cutting back on overproduction and excess inventories.

By increasing operational efficiency, automated guided vehicles (AGVs) and robotics reduce energy consumption. Reduced energy use and a smaller carbon impact result from fewer manual operations. IoT-powered climate and temperature control real-time energy use, humidity, and temperature are all tracked by IoT sensors. While preserving product quality, smart controls modify settings to reduce energy use. Reducing supply chains' environmental effect through the use of smart technologies is an essential first step toward sustainability. AI, blockchain, IoT, and predictive analytics can help businesses reduce emissions, eliminate waste, and improve operational efficiency.

To cut waste and increase product lifespans, the circular economy promotes recycling, reuse, and repair. Smart technologies allow businesses to transition from linear (produce-use-dispose) supply chains to circular ones. Businesses may cut packaging waste and utilize more sustainable materials with the use of AI and machine intelligence. By employing smart logistics, green warehousing, a circular economy models, sustainable packaging, and waste reduction strategies, businesses may meet their legal requirements, gain the trust of their customers, and contribute to a healthier planet.

11.4 Industry-specific Applications

Applications of SSCM in particular industries using smart technology show how companies can minimize their environmental impact while increasing productivity and profitability. In the manufacturing industry, real-time data and performance of machinery can be gathered by IoT sensors, and AI algorithms can thereafter examine the data. This will help in anticipating failures before their occurrence, cutting down on downtime and resource wastage, increasing equipment life, and optimizing machine usage to reduce energy consumption with the AI-powered energy management systems. A smart factory drastically improves labor efficiency, data privacy, quality control, and production (Kandarkar, 2024).

In the retail industry, just-in-time inventory replenishment is made possible through use of machine learning models that forecast demand trends accurately. This helps minimize waste and overproduction, reduce transportation-related emissions and storage expenses, and also ensures product availability, and thereby customer satisfaction. Application of blockchain technology ensures suppliers follow sustainability criteria through end-to-end–enabled product traceability, ultimately resulting in sustainable sourcing.

Considering the agriculture and food industry, by measuring crop health, weather, and soil moisture, smart sensors let farmers administer pesticides, fertiliser, and water more effectively. This results in minimized usage of chemicals and water, increased crop yield and reduced pollution thus leading to precision agriculture. Perishable items are monitored throughout storage and transit by IoT-enabled temperature and humidity sensors. Alternative food production and distribution schemes that rely on organizational or technological innovation to boost food production without negatively impacting the environment have emerged as a result of the sustainability challenge and the need to ensure the production of high-quality, reasonably priced, and healthful food. Sustainable food supply chains through smart farming are viewed as a promising approach. Both sustainable food supply chains and smart farming have the potential to improve farmers' incomes, reduce the negative environmental effects of agriculture, and create new kinds of value, and are typically seen as distinct routes to sustainability (Lioutas, 2020).

Likewise, in the automotive industry, AI can evaluate the condition of batteries and recommend best recycling or repurposing techniques. There are numerous advantages and effects to using IoT in SCM. IoT technology has therefore begun to be increasingly used in the automotive industry's SC (Pasi, 2020).

In the pharmaceutical industry, blockchain can help provide an immutable detail of the SC confirming the legitimacy of medications. This result in decreased use of fake medicines, improves patient safety, and increases trust in pharmaceutical products. In the energy and utilities sector, AI and smart meters use real-time supply and demand data to optimize energy distribution thus minimizing energy loss, encouraging utilization of sustainable energy sources and increased grid efficiency and reliability. AI can help optimize routes in logistics and transportation by analyzing traffic, delivery schedules, and weather. IoT-enabled telematics systems track both vehicle performance.

Smart technology advancements transformed to specific industries that promote environmentally friendly supply chain management are transforming company operations. Big data, blockchain, IoT, and AI all boost transparency, provide useful insights, and mitigate social and ecological effects across industries. By embracing such smart solutions, firms position themselves for long-term success while also improving supply chain robustness and supporting global ecological objectives.

11.5 Data and Risk Management

11.5.1 Data management

A sustainable futuristic supply chain management necessitates that supply chain operations address social, ecological, and commercial goals. These goals require competent risk and data management. Recent innovations such as artificial intelligence (AI), the Internet of Things (IoT), blockchain, and big data are redefining the way data is gathered, evaluated, and employed to minimize vulnerabilities and reinforce supply chains. Real-time information, predictive analytics, and secure data interchange can help businesses boost transparency, reduce their environmental impact, and better anticipate disruptions.

In SSCM, knowledge serves as a framework for decision-making and operational efficacy. Businesses can employ competent data collection, processing, and analysis to gauge sustainability KPIs, reveal inefficiencies, and evaluate success rates. Blockchain networks can help create an unchangeable and transparent record of material sourcing and transactions. ERP systems can combine information from logistics, production, inventory, and procurement. IoT devices can provide real-time operational data from sensors on manufacturing, storage, and delivery aspects. Market data and customer feedback can provide insight into how well a product performs and how sustainability is perceived.

Data types include operational data such as machine performance, energy usage; environmental data such as waste production, water usage, carbon emissions, etc; social data such as supplier compliance, labor practices, etc; and logistics information such as delivery schedules, shipping routes, and fuel usage.

Some of the data management strategies are as mentioned below:

- Data centralization and integration – AI-driven systems combine information from several sources into a unified dashboard. This helps offer a thorough understanding of SC while enhancing stakeholder collaboration.
- Accuracy of data and quality assurance – Machine learning algorithms are able to not only detect discrepancies in data but also automatically fix errors finally ensuring data integrity, increased accuracy of reports and projects, and improved adherence to legal standards.
- Privacy and data security – Use of blockchain encrypts and secures data that cannot be tampered with. This ensure prevention of illegal access and data breaches, guarantees adherence to data privacy laws such as GDPR, and therefore helps trust among the stakeholders.

11.5.2 Risk management

Operational disruptions, environmental difficulties, regulatory changes, and reputational harm are among the risks associated with SSCM. Risk identification, assessment, and mitigation are improved by smart technology. There are different types of risks such as environmental risks (pollution, natural disasters, etc.), operational risks (labor shortages, SC interruptions, equipment failure, etc.), regulatory risks (alterations in trade and environmental rules, etc.), and risks to reputation owing to unethical sourcing, environmental harm, and so on.

Predictive risk assessment – In order to forecast supply chain interruptions, AI models can accurately examine past data along with analysis of factors such as weather trends, geopolitical events, etc. This will reduce interruptions, minimize monetary losses while improving the time to react and adapt.Real-time monitoring – IoT sensors continuously monitor environmental variables, transportation conditions, and equipment performance. This will result in increased operational effectiveness, enhanced safety and quality of goods, and reduced waste and idle time.

Supply chain visibility – By logging each transaction and procedure, blockchain makes the SC transparent and auditable. It therefore, lowers the possibility of fake goods and unethical behavior; increases accountability of suppliers; and increases customer trust.

Scenario planning and simulation – With the help of real-time data and other external factors, AI models can prototype SC scenarios to minimize operational downtime and monetary losses and help enhance strategic planning for the long-term. For instance, AI models can evaluate how a new trade tariff may affect sourcing-related expenses and recommend substitute providers.

Ecosystem and ethical risk tracking – AI-powered systems examine the detrimental effect on information and supplier conduct, thus boosting the credibility of the brand and suppliers' conformance to sustainability requirements. Thus, data and risk management with smart technology can result in greater resistance and adjustment, higher environmental and social outcomes, cost savings, increased efficiency, improved compliance with legislation, and reinforced trust. Finally, developing an enduring and sustainable supply chain and securing long-term success in a competitive global market are significantly focused on data and risk management.

11.6 Conclusion

SSCM empowered by smart technology, has shifted corporate operations by increasing efficiency, lowering environmental impact while promoting greater social responsibility. Novel technologies such as artificial intelligence (AI), the Internet of Things (IoT), blockchain, and big data can assist firms by boosting visibility into their supply chains, avoiding interruptions, and maximizing resource utilization. Smart technology improves data quality, fosters real-time decision-making, and assures conformity to environmental and social norms.

Automation and predictive analytics contribute to the circular economy by lessening emissions, dipped waste, and advocating ethical work practices. Blockchain boosts supplier relationships and consumer trust by assuring transparency and obligation throughout the supply chain. The synthesis of smart technology and sustainability fosters long-term resilient organizations while also adhering to globally recognized ecological and regulatory requirements.

Smart technology in SSCM can help firms establish an appropriate equilibrium between sustainability and financial viability. Businesses can gain an edge over competitors by strategically limiting risks, strengthening processes, and stimulating transparency. Sustainable supply chains are no longer merely considered a competitive advantage, but rather a corporate need in the dynamically shifting global marketplace.

References

AlMulhim, A. F. (2021). Smart supply chain and firm performance: the role of digital technologies. *Business Process Management Journal*, *27*(5), 1353-1372.

Badr, B. (2022). Smart Technologies for a Sustainable Service Supply Chain: A Prospective Perspective. *Journal of Environmental Issues and Climate Change*, *1*(1), 77-89.

Büyüközkan, G., Göçer, F. (2018). Digital supply chain: literature review and a proposed framework for future research, *Comput. Ind.*, 97, 157-177

Golicic, S.L., Smith, C.D. (2013). A Meta-Analysis of Environmentally Sustainable Supply Chain Management Practices and Firm Performance. *J. Supply Chain Manag.* 49, 78–95.

Kandarkar, P. C., & Ravi, V. (2024). Investigating the impact of smart manufacturing and interconnected emerging technologies in building smarter supply chains. *Journal of Manufacturing Technology Management*, *35*(5), 984-1009.

Lee KL, Teong CX, Alzoubi HM, Alshurideh MT, Khatib ME, Al-Gharaibeh SM. (2024). Digital supply chain transformation: The role of smart technologies on operational performance in manufacturing industry. *International Journal of Engineering Business Management,* 16. doi:10.1177/18479790241234986

Lioutas, E. D., & Charatsari, C. (2020). Smart farming and short food supply chains: Are they compatible?. *Land use policy*, *94*, https://doi.org/10.1016/j.landusepol.2020.104541

Liu, L., Song, W., & Liu, Y. (2023). Leveraging digital capabilities toward a circular economy: Reinforcing sustainable supply chain management with Industry 4.0 technologies. *Computers & Industrial Engineering*, *178*, https://doi.org/10.1016/j.cie.2023.109113.

Liu, W., Long, S., & Wei, S. (2022). Correlation mechanism between smart technology and smart supply chain innovation performance: A multi-case study from China's companies with Physical Internet. *International*

Journal of Production Economics, *245*, https://doi.org/10.1016/j.ijpe.2021.108394.

Nasiri, M., Ukko, J., Saunila, M., & Rantala, T. (2020). Managing the digital supply chain: The role of smart technologies. Technovation, 96-97, https://doi.org/10.1016/j.technovation.2020.102121.

Pagell, M., Shevchenko, A. (2014). Why research in sustainable supply chain management should have no future. *J. Supply Chain Manag*. 50, 44–55.

Pasi, B. N., Mahajan, S. K., & Rane, S. B. (2020). Smart supply chain management: a perspective of industry 4.0. *Supply Chain Management*, *29*(5), 3016-3030.

Prause, G. (2019). Smart contracts for smart supply chains. *IFAC-PapersOnLine*, *52*(13), 2501-2506.

Rao, P., Holt, D. (2005). Do green supply chains lead to competitiveness and economic performance? *Int. J. Oper. Prod. Manag*. 25, 898–916.

Saryatmo MA and Sukhotu V. (2021). The influence of the digital supply chain on operational performance: a study of the food and beverage industry in Indonesia. *Sustainability*, 13(9). doi:10.3390/su13095109

Seuring, S., & Müller, M. (2008). From a literature review to a conceptual framework for sustainable supply chain management. *Journal of cleaner production*, *16*(15), 1699-1710.

Verma, A. (2024), "Green thinking: impact of smart technologies on supply chain management", Journal of Science and Technology Policy Management, Vol. ahead-of-print No. ahead-of-print. https://doi.org/10.1108/JSTPM-01-2024-0020

Yawar, S.A., Seuring, S. (2015). Management of Social Issues in Supply Chains: A Literature Review Exploring Social Issues, Actions and Performance Outcomes. *J. Bus. Ethics*, 1–23

Yousefi, S., & Tosarkani, B. M. (2022). An analytical approach for evaluating the impact of blockchain technology on sustainable supply chain performance. *International Journal of Production Economics*, *246*, https://doi.org/10.1016/j.ijpe.2022.108429.

https://www.sohu.com/a/400738529_120045186.

12

Role of Technology in Sustainable Logistics and Inventory Management with Reference to Food Corporation of India

Suresh Govindapuram[1], Chandrasekaran N.[1], Ramasubramaniam M.[2], and Rakesh Kumar[3]

[1]IFMR GSB, Krea University, India
[2]Area Chair – Business Analytics Professor of Operations & Analytics, Loyola Institute of Business Administration (LIBA), India
[3]Department of management study, Graphic Era Deemed to be University, India
E-mail: govindapuram.suresh@krea.edu.in

Abstract

Agriculture accounts for 18.2% of India's GDP, and more than 450 million people work there. Despite obstacles, supply chain integration (SCI) and developments in the agrifood industries are changing. SCI improves performance and guarantees food quality in the intricate global supply chain. SCI is converting food supply chains into incredibly effective digital systems by combining them with technological innovations like blockchain, artificial intelligence (AI), and the Internet of Things (IoT). AI is being integrated by the Food Corporation of India (FCI) to improve distribution, storage, procurement, and logistics. It helps increase control, coherence, and operational efficiency while tackling waste, delays, and inefficiencies. Furthermore, AI in agriculture is anticipated to transform food security and service delivery for stakeholders. This chapter explores different technological aspects that FCI uses and gives policy suggestions to improve the various processes.

Keywords: Supply chain, AI, FCI, food security, logistics, India.

DOI: 10.1201/9788743806905-12

12.1 Introduction

India has progressed far from being labeled as a purely agrarian economy, but agricultural and allied industries in India are still employing 42.3% of the population, which equals more than 450 million people, and is the biggest employment sector in India. The sector contributes around 18.2% to India's GDP (gross domestic product), reflecting a slight increase despite various challenges. The agrifood sector is highly significant in the development of the nation and has implications for the world's economy. Emerging growth brings increasing competition to the sector, and this induces novel solutions in the supply chain of food and food-related materials. The solutions push the food supply chain toward integration. Tan et al., 2017 highlighted the significance of supply chain integration (SCI) in ensuring food integrity amid the complex dynamics of supply chain activities at a global level. SCI involves the strategic synchronization and alignment of intra-administrative and inter-organizational processes, enhancing smooth coordination across the supply chain network. Those organizations that integrate their operations into external stakeholders systematically are likely to gain greater efficiency and better performance results (Frohlich & Westbrook, 2001). Besides the supply chain integration, increasingly the supply chains are prone to technological disruptions. Eckert et al. (2016) highlighted that Price Waterhouse Coopers (PWC) has eight disruptive technologies that would have changed the traditional supply chain into a digital supply chain: AI, IoT, augmented reality (AR), virtual reality (VR), robots, blockchain, 3D printing, and drones (Eckert et al., 2016). Thus, SCI along with adoption of new-age technologies such as AI is expected to give fillip to the sector. Studies by authors (Belhadi et al., 2021; McKinsey, 2018) anticipate a global GDP boost of USD 13 trillion through AI analytics by 2030 (approx. 16%), which is expected to disrupt different industries, particularly those in the supply chain like logistics and retailing. With widespread adoption of artifical intelligence (AI) technologies across industries, food supply chain is also embracing this new technology. AI is currently being considered to revolutionize the capability of the FCI in the logistics and supply chain management organizations. The adoption of AI technologies, therefore, is expected to boost this sector and contribute to a growth in market share. For example, according to MarketsandMarkets, the evaluated AI in agriculture market size is expected to grow at a rate of around 23.1% compound annual growth rate (CAGR) for the forecast period 2023–2028. One of the benefits of AI technologies is the supply chain operational efficiency (Khajavi & Holmström, 2015).

The efficiency of the food supply chain (FSC) can be improved using AI because it would give greater control over the supply chain and logistics (Kittipanya-Ngam & Tan, 2019). Given the pivotal role of logistics in the FCI's mission of ensuring food security, AI can help in ensuring food security by making the entire process more efficient through improvement of the FSC and coherence of the vast stakeholders. This would help FCI in addressing most of its challenges related to procurement, distribution, and storage of goods for better service delivery to farmers and consumers on a larger scale.

12.2 Background of FCI

Food Corporation of India (FCI), a public sector organization incorporated on January 14, 1965, by the Food Corporations Act of 1964, acts as the main executing agency for the Central Government's food policies.

India's economy is based on agriculture. Food security is essential for economic stability, as it increases agricultural productivity, resulting in more employment opportunities and better incomes for farmers. The National Food Security Act (NFSA) of 2013 seeks to supply subsidized food grains to two-thirds of India's population, focusing on vulnerable groups and other socio-economic backgrounds.

During the financial year 2023–2024, more than 1.27 crore farmers were benefited through the procurement of 1044 LMT (wheat + paddy) at a minimum support price (MSP) value of more than ₹2.27 lakh crore. The FCI is the key organization accountable for the implementation of the NFSA through the procurement, storage, and distribution of food grains. The FCI performs the core activities of the food grain supply chain, such as procurement, intra-state movement (within various regions of a state), inter-state distribution (from surplus to deficit states), and movement of food grains from district-level warehouses (DLWs) to destination fair price shops (FPSs), which are the final points of distribution. Public distribution system (PDS) forms the final link in this process, providing a channel for delivery of subsidized food grains from the regional warehouses to end-users, specifically those being categorized as below the poverty line (Mogale et al., 2019). A major part of the food supply chain (FSC) activity is carried out in partnership with other state agencies and authorities within India. In India, an integrated FSC network is established by carefully analyzing the nation's current food grain production, distribution, and purchase (Figure 12.1).

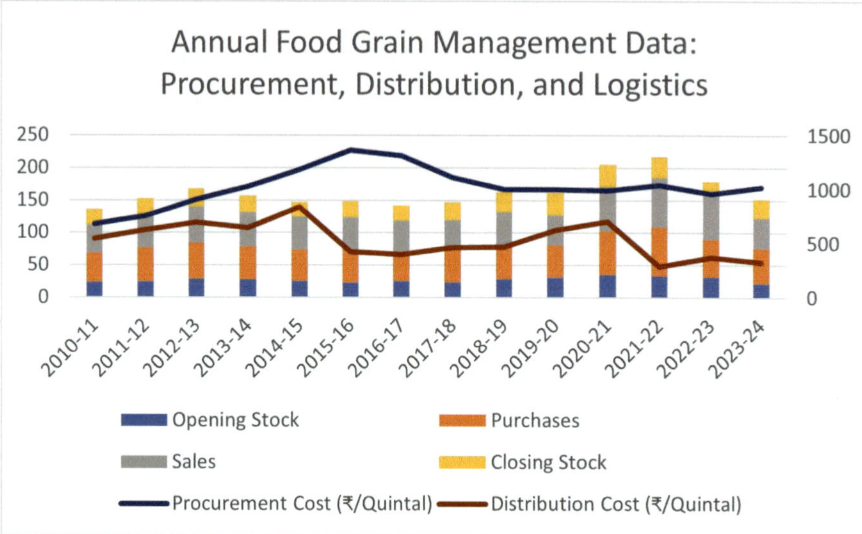

Figure 12.1 Trends in food grain stock, purchases, sales in million tons, procurement and distribution costs in ₹/Quintal (2010–2024).

Source: Annual Reports, FCI https://fci.gov.in/annualreport

Across the 14-year period, the graph highlights the purchasing and selling trends, storage facilities, and quantity transported. There is a remarkable consolidation in purchases following the last couple of years, more especially after 2020–21, with 2020 going even steeper as against 2021–2022, which could illustrate that strategized hiking in stock preparations might have been due to the government's initiatives in food security when Covid posed serious threats. Sales have also followed parallel trends with a hike in 2020–2021 and thereon.

It could be observed that changes in the distribution cost took place more heterogeneously compared to the procurement cost. The peak was observed during the year 2014–15, which is 835.23 crores, after which a significant drop took place during the year 2015–16. The distribution costs are seen to decline. This can be credited to the application of better measures in logistics.

All of this creates a favorable environment for disruptive technologies like artificial intelligence and data-driven systems, if they are applied to improve inventory control, optimize storage areas, and boost supply chain efficiency using tools. Closing these gaps would undoubtedly lead to alignment of sales, holding capacity, and procurements, removing any barriers to seamless operations.

12.2.1 Recent developments

For the fiscal year 2024–2025, PM Narendra Modi's Cabinet Committee on Economic Affairs (CCEA) has authorized an equity capital of Rs 10,700 crore for the FCI. As the government's foodgrain distribution and procurement agency, the FCI is essential to preserving strategic foodgrain supplies, distributing foodgrains for welfare programs, stabilizing market prices, and guaranteeing food security through procurement at MSP. Strengthening the agricultural industry and ensuring the welfare of farmers across the country are the goals of the CCEA ruling. From Rs 4,496 crore in the 2019–20 fiscal year to Rs 10,157 crore in 2023–24, the FCI's equity grew. The approval of this significant sum will improve the FCI's restructuring efforts and strengthen it financially.

The achievements of FCI include expansion of procurement to more than 60 times in the past 60 years. FCI is managing storage capacity to the tune of 410 LMT (owned + hired) in order to store this enormous amount of food grains. Approximately, 60 railway rakes carry 32 lakh bags daily on average. On average, each bag travels 1200 kilometers. Open market sale were implemented to keep the prices of rice and wheat stable. Every year, almost 80 crore recipients receive distributions of 600 LMT of food grains. FCI has roughly 27,000 employees and officers working under it, generating employment.

12.3 Need for AI in FCI Logistics

In accordance with the right to food, governments all over the globe consciously offer food subsidies to the poor in order to safeguard them from growing food costs, alleviate poverty, and provided that they get the best nutrition possible. Through tens of thousands of FPSs, the PDSs distribute subsidized food to 800 million people. Nonetheless, PDS's dysfunctional methodology and disparities among states have drawn criticism. Multiple stakeholders' engagement in PDS complicates information management and procurement, leading to economic inefficiencies and food waste. In many areas, caste, gender, power dynamics, and local politics all have an impact on food distribution due to a lack of transparency, which undermines public confidence (Pradhan & Rao, 2018). New technologies such as AI help in increasing awareness, provide an overseeing tool for effective distribution of subsidized goods (Bohtan et al., 2017), and combine the numerous PDS stakeholders to promote transparency (Bohtan et al., 2019).

Understanding the FCI supply chain is critical for tackling a number of issues. The increasing population coupled with post-harvest losses means higher production, procurement, and controlling the losses are required to ensure food availability. Although production has grown due to the application of sophisticated agricultural technology, policymakers are grappling with issues related to increasing procurement activities and controlling grain losses (Sharon et al., 2014). India is the world's second-largest producer of food grains, behind China (Khatkar et al., 2016) and still its agricultural regions lack adequate procurement hubs, thus limiting farmers' bargaining power. They struggle to gather goods and travel to distant centers to take advantage of minimum support price (MSP) auctions. This leads to lower local rates for selling goods to local dealers. The FCI struggles to aid farmers, and state government agencies store food crops in open places, causing deterioration in grain quality and increased losses.

Through advancements in AI adoption, innovation, and transformation in the provision of services, policy making, and enforcement potentially transform the public sector. An example provided by Montoya and Rivas (2019) is that AI may be used by governments to increase the quality of public services. Ojo et al. (2019) could add that AI emerges as sophisticated feature added in public administration to gain trust from citizens. Toll et al. (2019) contribute that AI enhances efficiency and effectiveness of public administrators through the appropriate application of the technology. Governments are able to use artificial intelligence (AI) to improve forecasting and modeling capabilities and to model different policy scenarios, thus allowing research with alternative policy choices (Margetts & Dorobantu, 2019). AI-based value creation extends across several areas of public administration, such as decision support, transportation, public health, and law enforcement (Gomes de Sousa et al., 2019). The use of AI in public governance can be potentially transformative by enhancing risk recognition, productivity, data processing, social interactions, and sustainable development (Zuiderwijk et al., 2021). There has been limited awareness of the extensive implications of AI in the workforce, organizational setups, economic frameworks, government processes, and overall society despite development in AI implementation in public services (De Sousa et al., 2019).

12.4 AI Applications in FCI Logistics

12.4.1 AI-driven operational and supply chain management systems

ANNA DARPAN (Integrated Supply Chain Management System) is a digital and efficient platform that reduces the time-consuming operation of FCI, automating the entire supply chain infrastructure process from start to finish. WINGS (Warehouse Inventory Network and Governing System) is a system which is responsible for the stacking of the operation, and the system uses GPS coordinates to find the mills and depots. This, in turn, helps to use that data to guide the advance route planning and finding out the distance optimization. CFPP (Centralized Food Grain Procurement Portal) may use AI or machine learning algorithms to analyze data, predict trends, and make strategic decisions. Its main role is promoting transparency and monitoring promising projects by integrating and overseeing data from various state procurement portals in real time. The portal has significant amounts of data that is related to farmer registration, procurement, and MSP transfers that are updated daily.

As of fiscal year 2023–24, FCI has been recognized for the successful reduction of the amount of food grains in FCI storage and during transportation. During the 2013–14 financial year, the combined net storage loss of wheat and rice was reported at 0.17%. By 2023–24, though, the Food Corporation of India (FCI) registered a net storage gain of 0.22%, showing notable progress in storage management and efficiency improvement. This change illustrates the success of strategic interventions in reducing post-procurement losses and maximizing the preservation of grains in the national food chain (it is not specifically mentioned that the reduction is due to AI, paragraph can be removed). Transit losses also showed a dramatic drop, falling from 0.46% in 2013–14 to 0.16% in 2023–24.

12.4.2 AI-based security and quality control systems

Its video-surveillance enclosures are monitored by AI technology when FCI secures its depots. Thus, any theft or attempt to damage salient grain stock can be caught at once. An automatic grain analyzer (AGA) is used with the integration of artificial intelligence technologies in the grain procurement system, ensuring that inspection for grain is accurate and transparent – alongside minimal human intervention. The AI system has quality control of wheat and rice concerning moisture and can identify below-grade products, which prohibit such products from entering the supply chain. The QMS software for quality control has been commandeered by the Department of Food and Public Distribution (DFPD) and FCI. The QMS software shows real-time values, including rejection rates, to ensure that high-quality standards are maintained throughout the distribution process.

12.4.3 AI-enhanced data and resource management

Vehicle location tracking system (VLTS) captures the real-time movement of trucks transporting food grains. The feasibility of integrating with an Android mobile app indicates the deployment of AI algorithms for the purposes of real-time route optimization and predictive analytics for transport logistics. The application of AI-driven auto-scaling for Oracle Cloud services implies use of AI to optimize the use of cloud resources. Similarly, AI methodologies could be integrated into the MIS applications so that PIMS (Procurement information management system) can analyze procurement data for predicting the volume of procurement at FCI, optimization of stock levels, and discovery of potential disruptions to the supply chain.

12.4.4 AI-driven customer interaction and support

Starting January 14, 2024, the FCI website features a chatbot powered by AI which gives customers quick answers to their questions and clarifies any doubts related to the organization's operations.

12.5 Areas of AI Integration in the Supply Chain

The integration of AI in food supply chain management can make the supply chain easier; facilitate big data analytics for risk assessment, and enables near real-time maximum decision-making through the analysis of huge amounts of data from diverse sources like the web, social media, history, and information systems of supply chain members. Emerging technologies like artificial intelligence can play a pivotal role in the formation of harmonized supply chains that thrive through collaborative sharing of knowledge and resources, which dictates organizational performance through Reduced OpEx and wastages. A proper logistics system is an effective form for alleviating bottlenecks because it covers most of the costs occurring in the value chain. Downstream supply chain management is in focus for optimal efficiency, sustainability, and competitiveness against market competition.

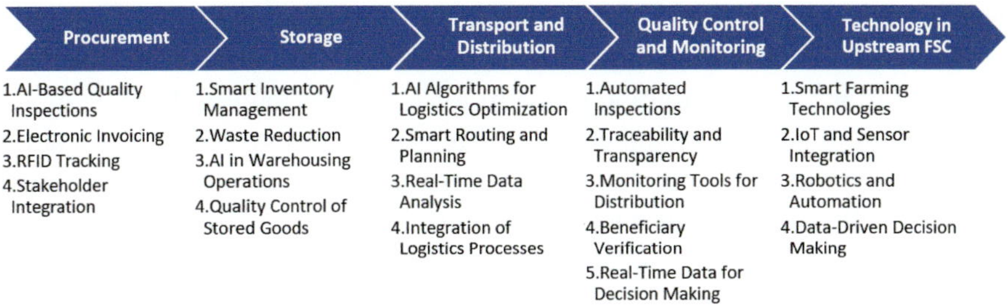

Figure 12.2 Food supply chain processes wherein AI can be applied.

The Indian think tank, NITI Aayog recommends using AI in PDS to improve the decision-making system, ensure quality service delivery, combat corruption, and improve transparency. AI can improve the PDS in its present form to ensure the sustainability of the existing PDS. Optimum acceptance and usage of intelligent technologies like AI is necessary in distributing services by remaining transparent in the distribution process, along with beneficiary authentication and the efficient administration of databases.

Algorithms, advanced data analytics, and AI can determine the effect of weather events or other supply chains shocks across the entire supply chain. Additionally, it can let one know if or when product emergencies are likely to occur (Morella et al., 2021). By integrating models of independent data analysis and repositories of past information with real-time data streams, integration with previous technologies enables more efficient and effective agri-food supply chain management (Wolfert et al., 2017).

The flowchart below (Figure 12.2) outlines the various applications of AI in different stages of the FSC processes, from procurement to transportation, storage, and quality control, with integration of advanced technologies across the system.

12.5.1 Procurement

AI-based quality assessments and electronic invoices can be used for sufficient procurement and packaging of food grains from food growers (Kumar et al., 2022). The PDS operates in a complex multi-stakeholder system involving the central government, state governments, the FCI, fair price shops, and beneficiaries. The elaborate governance structure presents serious challenges in the procurement, distribution, and management of foodgrain quantities as well as related information flows, which call for concerted policy interventions and administrative efficiency (Bohtan et al., 2020). This causes economic inefficiencies and waste (Chhotray et al., 2020). The bulk transfer of food grain from the central to the state government creates storage, transportation, and distribution problem for local distributors (Mogale et al., 2016). Digital allocation and control system measures issues of duplication before supplies enter ration shops (Masiero, 2020). The use of technology, such as RFID, to trace and track goods until delivered to the final destination will reduce item misplacement and loss (Biswal et al., 2019).

12.5.2 Storage

There is an increase in storage capacity of FCI throughout the years (Figure 12.3). A vast quantity of food grains (13,349 MT) during 2017–18 to 2021–22, had been reported destroyed due to poor maintenance practices. Of the total deteriorated stock, 14,980 metric tons of wheat worth 78% of the total losses deteriorated because of storage under the cover and plinth (CAP) process and caused a financial loss of ₹31.60 crore in Punjab as per an audit. AI can assist the actions of warehousing such as stock holding, overstocking, spoilage, and most of all the potential for optimum exploitation relating to warehouse activities. There exists empirical evidence that the implementation

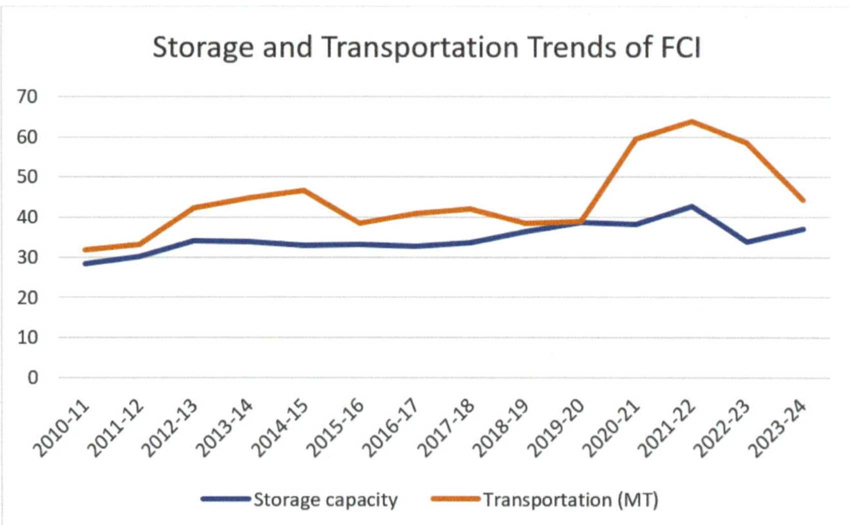

Figure 12.3 Storage capacity and transportation trends (in million tonnes) of food grains from 2010–24.

Source: Annual Reports, FCI https://fci.gov.in/annualreport

of AI in food chains would reduce wastage while preserving food quality and safety (Dora et al., 2022). AI can assist in enhancing the management of the storage processors. It can further help determine how warehouse activities would optimize the likelihood of improving storage efficiency and reduce food waste. AI can enable operational intelligence in inventory management in order to achieve waste reduction by predicting and pre-empting stock outs, overstocking, and spoilage.

An ever-evolving array of applications of AI is being developed in order to make it easier for people to perform their work (Chang et al., 2021). Such applications also have great potential in inbound logistics of food banks with regards to some expert knowledge-based input on the food quality of the donated food, data-processing methodologies, and processing-results feedback in order to enhance warehousing-management performance.

12.5.3 Transport and distribution

FCI moved a total of 443.46 LMT foodgrains during the FY 2023–24. Based on demand, weather, traffic, and inventory level, AI algorithms can optimize logistics and distribution by processing and analyzing large volumes of data. Various trackers are used, such as smart maps, route planning, and RFID trucks, all working together to enable the efficient transportation of food grain (Biswal et al., 2019). AI did show some contributions to an extent, beyond the environment of firms, on the ground of integration of logistics processes as a whole. BD & DM enables the AI to identify good locations from which to accrue maximum rationale and resource costs through the process of optimizing logisticians' operations, specifically distribution, transportation, and storage, but order and warehousing as well, to reduce their use of energy and other resources (e.g., labor, capital) based on better utilization and capacity scheduling (Krstić et al., 2022).

12.5.4 Quality control and monitoring

AI usage is projected to address food quality and safety concerns, improve transparency and traceability, and increase FSC efficiency (Toorajipour et al., 2020). Since the commencement of PDS, poor government policies and methods for tracking out misconduct have supported unscrupulous activities. AI has the capacity to validate beneficiaries, ensuring that only qualifying individuals will gain access to subsidized goods. This will enhance the verification process, reducing errors and fraud while making the distribution process more dependable. Another vital function is ensuring that decisions taken in the PDS system adhere to the law and scientific theory, avoiding manipulation or harm to people. This, in turn, keeps the fairness and legality of operations of the FCI down to the ground.

Another important application of AI is its ability to bring real-time data for speedy responses. This makes it possible for FCI to promptly respond to any changes in supply and demand conditions, thereby enhancing the agility of food distribution. Agri-food supply chain management witnesses a very progressive phase during which integrated intelligence transforms a situation from decision-making to decision-assigning, ultimately allowing a prediction of which decisions need to be undertaken. Several tasks executed by AI-powered robots and machine learning, diminish the reliance on human labor and alleviate the labor shortage through a reduction of human risk and on-the-job challenges. Using manual systems for the execution of movement and distribution reports have made FCI mechanical and laborious. An integration of AI systems like those into existing system such as the depot online system (DOS) can automate data generation and reporting, and decrease manual work. This would ease response times and enhance operations. This technology would thus be a big boost for FCI; conversely with use of AI, one is likely to lessen stressors on human strength and increase the efficiency of its employees (Ikram et al., 2024).

12.5.5 Technology in upstream FSC (Food production & farming)

ICT developments have enabled the design of food production systems that optimize productivity and reduce faults. In addition, advances in technology for smart farming, including sensors for the detection of diseases, to monitor meteorological conditions, and to estimate crop yields, are increasingly accessible (Kittipanya-Ngam & Tan, 2019). Agricultural supply chains are being called upon to incorporate technology in the form of the Internet of Things, blockchain, big data analytics, artificial intelligence, and robotics (Lezoche et al., 2020).

All of the aforementioned variables are influencing how the agriculture industry organizes and makes choices. AI-powered robots and machine learning can execute a variety of jobs that formerly required the knowledge of human resources, lowering dependency on manual labor and managing labor shortages, hence removing human risk. All of these contributions help to gain a competitive edge and capture economic value.

12.6 Conclusion

AI technologies have shown promise to improve supply chain efficiencies and this is evident through the FCI's adoption and the resultant benefits. AI has potential applications in various stages of the food supply chain. From integration with existing technologies like IoT, RFIDs, efficient tracking and inspection of both input and output materials, smart storage, efficient routing, and automation, AI brings in a widespread supply chain transparency and control. Thus, AI's integration into supply chain and logistics management significantly improves efficiency and effectiveness in operational strategies relating to functions like warehousing, transportation, procurement, and inventory management. From addressing inefficiencies in PDS to enhancing agricultural sustainability and fostering sustainable supply chain initiatives, AI offers significant supply chain process improvements in planning, supplier relationship, processing, distribution management, reverse logistics, and a wide array of applications. Such developments not only boost manufacturing, assembly, and process plants but also enable better public policy in areas like food security, public health, and host of other services. Future research and investments in AI-powered solutions can revolutionize supply chains, ensuring they are efficient, transparent, and sustainable.

References

Aayog, N. I. T. I. (2018). National strategy for artificial intelligence. Paper. June, 2019-01.

Ajay, Bohtan, K. Mathiyazhagan, and Prem Vrat. (2019). Modeling the public distribution system: A PO-P approach. *Opsearch, 56*, 1024-1066.

Belhadi, A., Mani, V., Kamble, S. S., Khan, S. a. R., & Verma, S. (2021). Artificial intelligence-driven innovation for enhancing supply chain resilience and performance under the effect of supply chain dynamism: an empirical investigation. *Annals of Operations Research, 333*(2–3), 627–652. bibhttps://doi.org/10.1007/s10479-021-03956-x

Biswal, A. K., Jenamani, M., & Kumar, S. K. (2019). The impact of RFID adoption on donor subsidy through for-profit and not-for-profit newsvendor: Implications for Indian Public Distribution system. *Socio-Economic Planning Sciences, 69*, 100687. bibhttps://doi.org/10.1016/j.seps.2019.02.002

Bohtan, A., Mathiyazhagan, K., & Vrat, P. (2020). Indian public distribution system in a supply chain perspective: a critical review and the way forward.

International Journal of Services and Operations Management, 35(3), 359. bibhttps://doi.org/10.1504/ijsom.2020.105376

Bohtan, A., Vrat, P., & Vij, A. K. (2017). Supply chain of the Indian public distribution system: a new paradigm. *Journal of Advances in Management Research, 14*(1), 110-123.

Carter, D. (2020). Regulation and ethics in artificial intelligence and machine learning technologies: Where are we now? Who is responsible? Can the information professional play a role? *Business Information Review, 37*(2), 60–68. bibhttps://doi.org/10.1177/0266382120923962

Chang, V., Gagnon, S., Valverde, R., & Ramachandran, M. (2021). Emerging trends and impacts of the rise of AI, data analytics and blockchain. *Journal of Enterprise Information Management, 34*(5), 1277-1286. bibhttps://doi.org/10.1108/JEIM-09-2021-555

Chhotray, V., Adhikari, A., & Bahuguna, V. (2020). The political prioritization of welfare in India: Comparing the public distribution system in Chhattisgarh and Jharkhand. *World Development, 128*, 104853. bibhttps://doi.org/10.1016/j.worlddev.2019.104853

Civin, L. (2024). The aspects of artificial intelligence in different phases of the food value and supply chain. www.academia.edu. bibhttps://www.academia.edu/105565915/The_Aspects_of_Artificial_Intelligence_in_Different_Phases_of_the_Food_Value_and_Supply_Chain

De Sousa, W. G., De Melo, E. R. P., De Souza Bermejo, P. H., Farias, R. a. S., & Gomes, A. O. (2019). How and where is artificial intelligence in the public sector going? A literature review and research agenda. *Government Information Quarterly, 36*(4), 101392. bibhttps://doi.org/10.1016/j.giq.2019.07.004

Dora, M., Wesana, J., Gellynck, X., Seth, N., Dey, B., & De Steur, H. (2020). Importance of sustainable operations in food loss: evidence from the Belgian food processing industry. *Annals of Operations Research, 290*, 47-72.

Eckert, V.H., Curran, C., & Bhardwaj, S.C. (2016). Tech breakthroughs megatrend: how to prepare for its impact. *Price Waterhouse Coopers.*

FCI Annual Reports bibhttps://fci.gov.in/annualreport

Frohlich, M.T., & Westbrook, R. (2001). Arcs of integration: an international study of supply chain strategies. *Journal of Operations Management, 19*(2), 185–200.

Haenlein, M., & Kaplan, A. (2019). A Brief History of Artificial Intelligence: On the Past, Present, and Future of Artificial Intelligence. *California Management Review, 61*, 5–14.

Ikram, A., Mehmood, H., Arshad, M. T., Rasheed, A., Noreen, S., & Gnedeka, K. T. (2024). Applications of artificial intelligence (AI) in managing food quality and ensuring global food security. *CyTA - Journal of Food, 22*(1). bibhttps://doi.org/10.1080/19476337.2024.2393287

Ismaeil, M. K. L., & Lalla, A. F. (2024). The role and impact of artificial intelligence on supply chain management: efficiency, challenges, and strategic implementation. *Journal of Ecohumanism, 3*(4), 89–106. bibhttps://doi.org/10.62754/joe.v3i4.3461

Kacprzyk, J., Lezoche, M., Panetto, H., Hernandez, J. E., & Del Mar Eva Alemany Díaz, M. (2020). Agri-food 4.0: A survey of the supply chains and technologies for the future agriculture. *Computers in Industry, 117*, 103187. bibhttps://doi.org/10.1016/j.compind.2020.103187

Khajavi, S.H., & Holmström, J. (2015). Manufacturing Digitalization and Its Effects on Production Planning and Control Practices. In *IFIP International Conference on Advances in Production Management Systems*, 179-185. Springer.

Kittipanya-Ngam, P., & Tan, K. H. (2019). A framework for food supply chain digitalization: lessons from Thailand. *Production Planning & Control, 31*(2–3), 158–172. bibhttps://doi.org/10.1080/09537287.2019.1631462

Krstić, M., Agnusdei, G. P., Miglietta, P. P., & Tadić, S. (2022). Evaluation of the smart reverse logistics development scenarios using a novel MCDM model. *Cleaner Environmental Systems, 7*, 100099. bibhttps://doi.org/10.1016/j.cesys.2022.100099

Kumar, R., Ganapathy, L., Gokhale, R., & Tiwari, M. K. (2022). Scaling up the digitalization of the public distribution system: A COVID-19 pandemic impetus. *IFAC-PapersOnLine, 55*(10), 55–60. bibhttps://doi.org/10.1016/j.ifacol.2022.09.367

Kumar, S., Raut, R. D., Queiroz, M. M., & Narkhede, B. E. (2021). Mapping the barriers of AI implementations in the public distribution system: The Indian experience. *Technology in Society, 67*, 101737. bibhttps://doi.org/10.1016/j.techsoc.2021.101737

Lezoche, M., Hernandez, J. E., Del Mar Eva Alemany Díaz, M., Panetto, H., & Kacprzyk, J. (2020). Agri-food 4.0: A survey of the supply chains and technologies for the future agriculture. *Computers in Industry, 117*, 103187. bibhttps://doi.org/10.1016/j.compind.2020.103187

Margetts, H., & Dorobantu, C. (2019). Rethink government with AI. *Nature, 568*(7751), 163-165.

Masiero, S. (2020). Biometric infrastructures and the Indian public distribution system. *South Asia Multidisciplinary Academic Journal, 23*. bibhttps://doi.org/10.4000/samaj.6459

Margetts, H., & Dorobantu, C. (2019). Rethink government with AI. *Nature, 568*(7751), 163-165.

Mogale, D., Kumar, S. K., Márquez, F. P. G., & Tiwari, M. K. (2016). Bulk wheat transportation and storage problem of public distribution system. *Computers & Industrial Engineering, 104*, 80–97. bibhttps://doi.org/10.1016/j.cie.2016.12.027

Mogale, D. G., Ghadge, A., Kumar, S. K., & Tiwari, M. K. (2019). Modelling supply chain network for procurement of food grains in India. *International Journal of Production Research, 58*(21), 6493–6512. bibhttps://doi.org/10.1080/00207543.2019.1682707

Morella, P., Lambán, M. P., Royo, J., & Sánchez, J. C. (2021). Study and analysis of the implementation of 4.0 technologies in the Agri-Food Supply Chain: a state of the art. *Agronomy, 11*(12), 2526. bibhttps://doi.org/10.3390/agronomy11122526

Olsen, T.L., & Tomlin, B. (2019). Industry 4.0: Opportunities and Challenges for Operations Management. In *Tuck School of Business Working Paper No. 3365733*, The University of Auckland Business School Research Paper; The University of Auckland Business School: Auckland, New Zealand.

Ojo, A., Mellouli, S., & Ahmadi Zeleti, F. (2019, June). A realist perspective on AI-era public management. In *Proceedings of the 20th Annual International Conference on Digital Government Research* (pp. 159-170).

Sharifmousavi, M., Kayvanfar, V., & Baldacci, R. (2024). Distributed Artificial Intelligence application in Agri-food Supply chains 4.0. *Procedia Computer Science, 232*, 211–220. bibhttps://doi.org/10.1016/j.procs.2024.01.021

Strusani, D., & Houngbonon, G.V. (2019). The Role of Artificial Intelligence in Supporting Development in Emerging Markets. *EMCompass, 69*.

Tan, K.H., Ali, M., Makhbul, M., Ismail, A. (2017). The impact of external integration on halal food integrity *Supply Chain Management: an International Journal, 22*(2), 186-199.

Toorajipour, R., Sohrabpour, V., Nazarpour, A., Oghazi, P., & Fischl, M. (2020). Artificial intelligence in supply chain management: A systematic literature review. *Journal of Business Research, 122*, 502–517. bibhttps://doi.org/10.1016/j.jbusres.2020.09.009

Wolfert, S., Ge, L., Verdouw, C., & Bogaardt, M. (2017). Big Data in Smart Farming – A review. *Agricultural Systems, 153*, 69–80. bibhttps://doi.org/10.1016/j.agsy.2017.01.023

Zuiderwijk, A., Chen, Y., & Salem, F. (2021). Implications of the use of artificial intelligence in public governance: A systematic literature review and a research agenda. *Government Information Quarterly, 38*(3), 101577. bibhttps://doi.org/10.1016/j.giq.2021.101577

13

Identifying Key Food Values Influencing Sustainable Purchase Intentions: An EFA-Based Market Segmentation Approach

Madhusmita Choudhury[1], Shantanu Raj[2], and AdityaRanjan Samal[2]

[1]Department of Marketing, GITAM deemed to be university, India
[2]Centre of AgriManagement, Utkal University, India
E-mail: shantanuraj022@gmail.com

Abstract

Selling generic food items versus organic food items has many challenges everywhere unless the customer understands the underlying health benefits; it is like learning science; the more you know, the better you will be with the application. Similarly, understanding what a consumer is looking for in sustainable food and knowing how to profile the product for marketers is crucial for the product success in developing economies. This assumption served as the foundation for the current study's goal of addressing the gap left by the previous research. Among these are the food values that consumers place the highest priority on when making purchases of sustainable food items. To explore a range of food values associated with sustainable food orientation, this study divided the consumers into three distinct customer profiles. The Lusk & Briggeman food dimensions framework was employed to assess consumers' intentions to buy sustainable foods in underdeveloped nations.

Keywords: Sustainable food, health conscious, market segmentation, customer profiling, theory of planned behavior.

13.1 Introduction

In today's world, consumers are changing their food preferences toward sustainable food systems. In developed countries, sustainability in the food

DOI: 10.1201/9788743806905-13

sector is a significant global concern (Piracci et al., 2023), and it has become a central focus in the United Nations' Agenda (Reisch et al., 2013; Purvis et al., 2019). Western diets, in particular, are shaped by sustainability concerns such as ethical dilemmas, animal welfare, and environmental impact (Aleksandrowicz et al., 2016; Carlsson-Kanyama & Gonzalez, 2009). The sustainable food and agri-food sectors work together to establish a food system that is socially just, economically feasible, and ecologically sound, ensuring a resilient future. Sustainability in agri-food products has become a key topic of discussion (Borsellino et al., 2020; El et al., 2021), and sustainable food choices directly influence consumer behavior (Borsellino et al., 2020).

Food choices are complex and shaped by various factors such as sociodemographic factors, culture, norms, values, attitudes, and consumption habits (Furst et al., 1996). This complexity has drawn researchers to explore the motivations behind sustainable food choices (De Boer et al., 2007). The term "food motives" refers to the reasons or motivations behind consumers' food choices, and its significance has been highlighted in the context of sustainable food consumption (Verain et al., 2016). Factors like socioeconomic and environmental benefits drive some sustainable food choices, but barriers such as reduced taste and convenience can hinder them. Sustainable food consumption also considers CO_2 emissions, water use, food security, food safety, animal welfare, seasonal and local production, and moral working conditions (Aiking & de Boer 2004).

These are the key factors influencing sustainable food choices, and extensive research has been conducted in developed countries on these topics. Some studies have suggested that promoting local and seasonal consumption supports local economies and fosters diversification (Aleksandrowicz et al., 2016). Consumers often associate sustainability with taste and nutrition (Barone et al. 2020), but sustainability encompasses many dimensions, making it challenging to provide a precise definition. Recent research has shown that animal welfare, food safety, and ethical considerations are crucial, with consumers in developed nations expressing a strong interest in animal welfare when purchasing food products (Bazzani et al. 2018).

13.1.1 Conceptualization of sustainability and sustainable food consumption

Sustainability is a multifaceted concept and constitutes several dimensions. Hence, there is no binding definition of sustainability. To comprehend sustainability, consumers need to possess integrated knowledge of various aspects of sustainability (Watson et al., 2010; Sanchez–Chapparo et al., 2022). Consumers process numerous sustainability-related information (Leonidou, C. N., & Skarmeas, D., 2017). According to the definition of the Food & Agriculture Organization (2010), sustainability encompassed "diet protective, preserving biodiversity, culturally acceptable, accessible, economically fair, affordability, nutritionally adequate, safe and healthy while optimizing natural and human resources" (Burlingame & Dernini, 2010). Sustainable food should adhere to environmental, political, and safety standards (Lu, Y. et al.; Alam et al., 2020), such as providing everyone with a safe, wholesome meal (Alam et al., 2020; Varzakas & Smaoui, 2024), and is a viable livelihood for farmers (Raghavendra et al., 2024), safeguard biodiversity, and generate minimum waste (Kumar et al., 2024). Consumers may promote social fairness, economic viability, and environmental health by selecting meals produced with as little influence on the environment as possible. This strategy encourages a robust and just food system for upcoming generations while supporting farmers and resource conservation. Customers may demand ecologically friendly practices by prioritizing sustainable food choices, ultimately leading to a more just and resilient food system for coming generations (Vermeir et al., 2020).

In the current economic and social context, it is highly required and urgent to understand the consumers' attitudes toward purchasing sustainable food products and identify the pertinent dimension in the Indian context that drives consumers toward sustainable food products. This assumption served as the foundation for the current study's goal of addressing the gap left by the previous research. Among these are the food qualities that consumers place the highest priority on when buying sustainable food items. *The researcher wants to explore which food values trigger sustainable purchase intention through EFA techniques and identify the market segment based on the priorities of the food values driving sustainable food choices.* To the author's best understanding and knowledge, the study is the first attempt to explore the relevant factor and create the cluster based on the food values posited by Lusk & Briggeman (2009).

13.1.1.1 Food values

Consumer preferences and habits are very dynamic and change from time to time. This is due to changes in consumer needs, preferences, and knowledge. Nevertheless, the underlying food values governing food preferences are relatively stable (Huang et al., 2024). Lusk & Briggeman (2009) pioneered the concept of food values. They explored 11 food values, which can be

Table 13.1 Major research in sustainability and sustainable food consumption.

Study	Research focus	Research type
Food sustainability: problems, perspectives and solutions	This study examines several viewpoints on attaining food sustainability and addresses the environmental effects of the global food system, such as greenhouse gas emissions, biodiversity loss, and water pollution.	Conceptual paper, UK
Effectiveness, challenges, and environmental impacts of new food strategies with plant and animal protein products	This study divulged the pertinent factor that impacted on the environment of novel food strategies utilizing plant and animal proteins, emphasizing the necessity of sustainable food production methods.	Empirical paper, Brazil
Sustainability and the food industry: A bibliometric analysis	This study examined the most often referenced works, prestigious journals, and potential avenues for future research on sustainability in the food sector.	Review paper, Turkiye
The role of food industries in sustainability transition: a review Factors that facilitate consumer uptake of sustainable dietary patterns in western countries: a scoping review	An overarching research on the food industry's sustainability transition, focusing on altering customer preferences to sustainable and healthy diets.	Review paper, India
Factors that facilitate consumer uptake of sustainable dietary patterns in western countries: a scoping review.	This study identified the salient factor that motivated consumers to embrace and adopt environmentally sustainable diets using scoping review techniques.	Empirical paper, US

Source: Author's Data Analysis

seen as the intermediary values or end state of existence linked with food purchase and consumption. Despite cultural differences in region and place, food values play a catalytic role in purchasing food products in grocery stores or organized retail outlets. Hence, food values effectively elucidate consumer dietary habits to inform marketing choices (Piracci et al., 2023; Daviet & Nave, 2024). In addition, the food values investigate the food choice preferences irrespective of food category because it encompassed general constructs, captured the various dimensions of food consumption, and espoused more specific product attributes (Bazzani et al., 2018). The scale has been widely applied in numerous areas predicting consumer behavior in organic food (Lusk, 2011), functional food (Pappalardo et al., 2016), and food nanotechnology (Yang & Hobbs, 2020), in addition to tracking alterations in dietary habits throughout the COVID-19 outbreak (Cerroni et al., 2022). The current study explores how food values influence customers' decisions to buy sustainable food items. Table 13.1 showing important study related with sustainable food consumption.

13.2 Theoretical Framework

Several theories have been proffered in the literature to expatiate the sustainability of food products. For instance, few studies have applied VBN (value-belief norm) to investigate the impact of human values on the behavior of environmentalists. TPB explains the relationship between behaviors, values, beliefs, and norms (Chen, 2015; Alam et al., 2020). The VBN model was an integral part of many studies. The theory of planned behavior, or TPB (Huynh, 2018), is the most widely applicable theory among the various theories put forth in the research to explain consumer behavior. Consequently, this study will use the TPB to forecast which important aspects affect customers' intentions to buy sustainable food products. The extant studies supported the idea that TPB can explain and predict the broader gamut of human behaviors (Rhodes & Yao, 2015; Alam et al., 2020).

The TPB is the extension of the TRA, which can examine human behavior's intricate traits. The TPB entails attitude playing a critical role in investigating the intention. The social norms state the individual's perception toward social pressure and how easy and challenging to perform a specific task is connoted by perceived behavioral control.

In brevity, the TPB explicitly provides plausible reasons and explanations of why specific consumers select certain products (Hassan et al., 2016; Mhlophe & Zulu, 2022). Attitude, subjective norms, and perceived behavioral control (Ip et al., 2022) can explain 40–50% of the variance in anticipating consumer intention. These factors can also account for 19–38% of the variance in behavior (Sun et al., 2020; Alam et al., 2020). The conspicuous difference between TPB and other theoretical frameworks is the amount of variance explained by TPB, which no other model has yet explained; hence, this model has ample applicability in predicting consumers' buying behavior.

Numerous research studies on food consumption-related behavior have applied TPB as their theoretical framework (Sparks et al., 1995; Conner & Sparks, 2005; Honkanen et al., 2005; Bonne et al., 2007). The TPB has a broader application in the organic food buying context (Cook et al., 2000;

Magnusson et al., 2001; Tarkiainen & Sundqvist, 2005) and in online purchasing behavior (Mathieson, 1991; Harrison et al., 1997; George, 2002; Bhattacharjee, A., 2002; Pavlou, 2002, August; Suh & Han, 2003). Although TPB has been widely applied in food consumption-related studies, only a meager amount of studies have stressed the uses of TPB as a theoretical framework to explore sustainable food consumption behavior. In their research paper, Vermeir & Verbeke (2008) have delved into the impact of attitude, social norms, and perceived behavioral control on sustainable food consumption. Hence, more research is needed to apply TPB's influencing factors to delineate the course of sustainable food product consumption behavior.

13.3 Materials and Methods

13.3.1 Instrument development

We administered the Lusk & Briggeman (2009) adopted study scale to test the questionnaire's validity and reliability. Most of the questions used to measure the latent components were taken from previous studies and adjusted for the specifics of this investigation. After subject matter experts from academia and administration reviewed the survey instrument, a pilot survey was carried out, and the validity and reliability of the questionnaire were tested using the 93 responses received during the study. By analyzing the pilot data, it was possible to determine that the questionnaire was reliable because of the

Table 13.2 Questionnaire items and corresponding codes were used in the study

NO.	Codes	Items
1	V1	Packed food is produced without adding chemicals to it.
2	V2	Packed food is produced using modern technology.
3	V3	Packed foods are good to taste and senses.
4	V4	Price has a direct relationship with packed food purchases.
5	V5	Consumption of packed food will not cause illness.
6	V6	Packed foods which are easy to cook saves time and energy.
7	V7	Packed foods contain a balanced amount of fats, proteins, and vitamins; in it.
8	V8	The quality of packed food is determined based on its place of origin.
9	V9	The production and marketing of packaged food benefited all stakeholders equally.
10	V10	Packed foods look very appealing.
11	V11	There is a positive impact of modern-packed food on the environment.
12	V12	Packed food promotes traditional pattern of consumption.

Source: Adpated from the study of Lusk, J. L., & Briggeman, B. C. (2009). Food values. American journal of agricultural economics, 91(1), 184-196.

composite reliability and Cronbach's alpha values, which were higher than (Martins et al., 2023) the 0.7 criterion (see Table 13.2).

13.3.2 Data collection and sample

The questionnaire is framed to investigate the consumers' behavior toward sustainable food choices. Bhubaneswar City is selected for research purposes due to the spurt of many sustainable food brands within the short span of five years. On the other hand, the city is rich in culture and diversity. It fosters local or traditional products which pursue sustainability in the food supply chain both environmentally and economically. Questionnaires were used to gather data between July 2024 and August 2024 through Google Forms for this study from the faculties of an Indian metropolis. This population stratum was selected based on their familiarity with the Internet and willingness to adopt sustainable products. Purposive sampling is used to gather quality data (Tongco, 2007). Participants below the range of 18 have been excluded from the study. The researcher assumed that with a minimum graduation degree, one could harbor adequate knowledge about sustainability.

For the screening of participants, we have inducted two questions: "Are you aware of Sustainability & possess some information about this concept?" and the second question is "Are you aware of any sustainable food brand in the market?" After receiving the positive responses to the question above, we forwarded the questionnaire to 215 participants and deemed it suitable for analysis in 151. Questionnaires were used to gather data from the faculties of an Indian metropolis for accomplishing the research work.

Table 13.3 presents the socio-demographic characteristics of the sample group, including age, educational level, family type, family size, occupation, and income level. The sample contained (43.7%) age group of 21–30, the highest sample percentage from the other age category. Most respondents are postgraduates (60.9%) and belong to the nuclear family (64.2%). Most of the sample had more than five members in their house (66.8%). Most of the respondents belong to more than 8 lakh income bracket (49%), with respondent occupations as private services (60.9%).

All the questions are summated in Likert scale. The sample analysis involved a descriptive analysis of the respondents, including age, income, family members, occupation, and types of family. The EFA and cluster analysis was done to segment the consumer based on demographics and their perceptions of sustainable food products. The analysis is facilitated by statistic software "SPSS (24 version)" and "Jamovi."

Table 13.3 Description of 151 used data for research purposes

Demographics		Count	Percentage (%)
Age	21–30	66	43.7
	30–39	42	27.8
	39–48	21	13.9
	48–57	17	11.2
	57–66	3	1.9
	>66	2	1.3
Education	Graduate	11	7.2
	Post graduate	92	60.9
	Others	48	31.7
Family	Joint	54	35.7
	Nuclear	97	64.2
Members	1–3	47	31.1
	3–5	55	36.4
	>5	101	66.8
Occupation	Govt servant	18	11.9
	Private	92	60.9
	Others	41	27.1
Income (In lakhs)	2–5	51	33.7
	5–8	27	17.8
	>8	74	49.0

Source: Primary data-researcher's work & authors' calculations.

13.4 Data Analysis

13.4.1 Exploratory factor analysis

We employed the exploratory factor analysis process to identify the unobserved components, primary drivers for consideration, and study which food values activate sustainable food purchase intention by EFA techniques.

13.4.1.1 Reliability analysis and internal consistency of the data

Diagnostic tests were performed using Cronbach's Alpha test (Hair et al., 2010) and the KMO test proposed by Kaiser (1974) to assess the internal consistency and sample adequacy across all items or replies from the questionnaire survey, which is 0.755, i.e., the value is closer to 1. Cronbach's alpha test was utilized to test the internal consistency of each factor. The relationships are good because their Cronbach's values are more significant than 0.9. The KMO value of EFA is 0.755, and the *p*-value for Bartlett's test of sphericity is <0.05. Our study's minimum threshold criterion for EFA is met. To eliminate misspecification bias and untrustworthy inference, items were assessed before moving on to the factor analysis. Communities demonstrate how the extracted factors have considered a variable's variance. Because of the low KMO value, we have eliminated one variable: "Packed food is produced using modern technology." The author has identified four factors: perceived health benefits, environmental concern, price benefits, and hedonism attributes after performing EFA (see Table 13.4). We observed that the price benefit factor becomes the better indicator of purchased intention toward sustainable food products, with the highest mean and lowest SD also comparatively less, indicating less variability in the dataset.

Table 13.4 EFA results

NO.	Items	Factor 1	Factor 2	Factor 3	Factor 4	Communalities
		Perceive health bene-fits	**Environmental concern**	**Price benefits**	**Hedonism**	
V1	Packed food is produced without adding chemicals to it	**0.739**				**0.615**
V3	The consumption of packed foods is appealing to taste and senses				**0.832**	**0.700**
V4	Price has a direct relationship with packed food purchases			**0.881**		**0.816**
V5	Consumption of packed food will not cause illness	**0.823**				**0.708**
V6	Packed foods which are easy to cook saves time and energy			**0.759**		**0.719**
V7	Packed foods contain a balanced amount of fats, proteins, and vitamins; in it	**0.76**				**0.642**
V8	The quality of packed food is determined based on its place of origin	**0.583**				**0.581**
V9	The production and marketing of packaged food benefited all stakeholders equally		**0.509**			**0.688**
V10	Packed foods look very appealing				**0.751**	**0.607**

Table 13.4 *Continued.*

NO.	Items	Factor 1	Factor 2	Factor 3	Factor 4	Communalities
		Perceive health bene-fits	**Environmental concern**	**Price benefits**	**Hedonism**	
V11	Modern-packed food has a positive impact on the environment		**0.809**			**0.673**
V12	Packed food promotes traditional consumption pattern		**0.772**			**0.626**
	Factor mean	**2.4**	**2.6**	**4.0**	**3.6**	
	Eigen value explained (%)	**25.3**	**16**	**12.6**	**9.3**	
	Factor SD	**.81084**	**.79668**	**.68312**	**.86923**	

Source: Primary data-researcher's work & authors' calculations.

13.4.2 Segmentation analysis

While factor analysis summarizes dimension at a macro element, cluster analysis allows the grouping of homogenous objects. *K* means cluster analysis has been used. Three clusters have been identified, with 59, 48, and 44 cases in each cluster, and all are relatively balanced. Table 13.5 helps us understand that 59 customers are in the cluster 1 segment, which is the highest. Based on information derived AIC, BIC, CAIC, and entropy within the cluster (Castejón J et al., 2016; Chakraburty et al., 2021). From Table 13.5, it is evident that cluster 3 is the ideal cluster solution for the study, indicating a better fit and more distinct clusters.

Table 13.6 depicts the ANOVA, and all the variables taken for the study are statistically significant because of the condition – p-value $<.05$. Hence, all the variables significantly contribute to cluster formation. We have applied *K* mean cluster analysis. Table 13.6 exhibits the final cluster center denoting the cluster's mean. Cluster is named based on the mean value.

First cluster has 59 respondents, followed by cluster 2, having 48 respondents, and the third constitutes 44 respondents. The first consumer class

Table 13.5 Number of cases in each cluster and class details

Class	Log-likelihood	AIC	CAIC	BIC	Entropy
1	–1796	782	1116	1011	0.966
2	–1718	721	1007	998	0.918
3	**–1667**	**716**	**1002**	**997**	**0.9**
4	–1598	825	1141	1256	0.96

Source: Primary data-researcher's work & authors' calculations.

(39%) scored highest on (v1, v2, and v8) variables. They are assigned as "Health driven." They seek a food product devoid of additives and chemicals, safely produced through modern technology, whose place of origin signifies the quality of ingredients used in food products. The second consumer class (31%) gave more priority to variables (v3, v4, v6, and v10). They are named as "Pleasure seekers." They seek sensory appeal in food and convenience. They are price-conscious people. The third consumer class (29%) gives more relevance to the variables (v5, v7, v9, v11, and v12). They are labeled as "Sustainability focused." They value food safety and balanced nutrition, and every actor involved in processed food production must receive a fair share of economic benefits. They are the leading exponent of environmental benefit.

Table 13.6 Cluster center and ANOVA table

Items	Variables	Clusters			ANOVA table			
		11	2	3	Mean square	df	F	Sig.
V1	Packed food is produced without adding chemicals to it	**4.2**	2.9	1.3	27.681	2	28.232	.000
V2	Packed food is produced using modern technology	**4.4**	3.0	4.0	1.402	2	2.855	.046
V3	The consumption of packed foods is appealing to taste and senses	2.5	**4.2**	3.5	1.467	2	1.239	.293
V4	Price has a direct relationship with packed food purchases	2.2	**4.0**	2.6	21.087	2	40.759	.000
V5	Consumption of packed food will not cause illness	2.1	2.3	**4.5**	3.691	2	3.901	.023
V6	Packed foods which are easy to cook saves time and energy	4.0	**4.6**	3.1	13.174	2	24.293	.000
V7	Packed foods contain a balanced amount of fats, proteins, and vitamins; in it	3.6	2.8	**4.2**	5.601	2	5.793	.004
V8	The quality of packed food is determined based on its place of origin	**4.1**	3.3	1.8	24.151	2	25.683	.000
V9	The production and marketing of packaged food benefited all stakeholders equally	3.6	3.2	**4.2**	4.267	2	4.280	.016
V10	Packed foods look very appealing	2.0	**4.5**	3.9	3.749	2	3.905	.023
V11	Modern-packed food has a positive impact on the environment	2.0	3.6	**4.0**	32.200	2	45.709	.000
V12	Packed food promotes traditional consumption pattern	1.8	3.0	**4.1**	21.416	2	24.847	.000

Source: Primary data-researcher's work & authors' calculations.

13.4.3 Segment profiling

The above table and figure discussed about the health-driven, Pleasure Seeker and group focused on sustainability relating the profiling variable.

13.4.3.1 Health-driven

Health-conscious packaged food items are natural, and no toxic chemicals are used during preparation. Vacuum sealing, flash freezing, or cold-pressed processes retain freshness without sacrificing safety. These products use modern technology to maintain nutritional integrity and shelf life (Rupp, 2003). The meal's location can also impact its quality because foods from areas with excellent agricultural practices or distinctive culinary traditions tend to have better flavor, purity, and nutritional value. Health-driven packaged goods provide consumers with a safer, higher-quality option by combining cutting-edge techniques and thoughtful sourcing.

13.4.3.2 Pleasure seeker

Packed meals have a tasty and fulfilling sensory experience that appeals to pleasure-seeking consumers (Locher et al., 2005). The food presentation and packaging attractiveness create a sense of quality and desirability (Van Esch et al., 2019). Price is also essential since satisfied customers are more likely to buy packaged meals that fit their budgets (Ragaert et al., 2004). Furthermore, packed foods appeal primarily to people with hectic schedules who still want to enjoy delicious and savory meals because of their convenience, especially with quick-cook choices that save time and energy (Raj et al., 2024).

13.4.3.3 Sustainability focused

Sustainability-focused packed meals are intended to be health-friendly, avoiding sickness through safe, high-quality ingredients. These products support a healthy diet because they are made with necessary nutrients, including lipids, proteins, and vitamins, in a balanced mix. A just and moral supply chain is created by designing the production and marketing procedures so that farmers, manufacturers, and consumers all gain equally from it. Additionally, eco-friendly materials and production techniques are used in modern packaging technologies to reduce their adverse effects on the environment. These products protect traditional eating habits and advance ecological health using sustainable practices while supporting traditional consumption patterns (Dubreuil, 2018).

13.5 Conclusion

Motivated by health, the fundamental concern of consumers is safe, natural production methods that do not involve hazardous chemicals, guaranteeing that the food is nutritious and wholesome. While preserving nutritional content, shelf life can be increased using vacuum sealing, flash freezing, and cold-pressing techniques. The product's point of origin also affects its quality because some areas may have superior farming methods or traditional food preparation methods that improve the flavor, nutritional value, and purity of packaged goods. Price can affect decisions about what to buy, even while you are having a great meal. Since these items are designed to accommodate busy lifestyles, the ease of preparation of the available options adds even more allure. To cater to the demands and cravings of customers who seek pleasure, marketing tactics should emphasize sensory experiences, cost-effective products, and convenience. Marketing efforts for the pleasure-seeking category should focus more on sustainable ingredients and traditional consumption patterns and promote the same to the target audience, appealing to their desire for health and environmental responsibility. Despite their disparate priorities, the quality, safety, and responsible production requirements unite these three consumer sectors. Businesses stand to gain by catering to the demands of all three sectors through integrated marketing communication (IMC), which blend pleasure, health, and sustainability components to create messages that appeal to various consumer demographics.

Acknowledgment

Thanks to the conference organizer upon providing an opportunity to do this research. Usual disclaimers apply.

References

Aiking, H., & De Boer, J. (2004). Food sustainability: Diverging interpretations. British Food Journal, 106(5), 359-365.

Alam, S. S., Ahmad, M., Ho, Y. H., Omar, N. A., & Lin, C. Y. (2020). Applying an extended theory of planned behavior to sustainable food consumption. Sustainability, 12(20), 8394.

Aleksandrowicz, L., Green, R., Joy, E. J., Smith, P., & Haines, A. (2016). The impacts of dietary change on greenhouse gas emissions, land use, water use, and health: a systematic review. PloS one, 11(11), e0165797.

Barone, B., Rodrigues, H., Nogueira, R. M., Guimarães, K. R. L. S. L. D. Q., & Behrens, J. H. (2020). What about sustainability? Understanding consumers' conceptual representations through free word association. International Journal of Consumer Studies, 44(1), 44-52.

Bazzani, C., Gustavsen, G. W., NaygaJr, R. M., &Rickertsen, K. (2018). A comparative study of food values between the United States and Norway. European Review of Agricultural Economics, 45(2), 239-272.

Bhattacherjee, A. (2002). Individual trust in online firms: Scale development and initial test. Journal of management information systems, 19(1), 211-241.

Blashfield, R. K., & Aldenderfer, M. S. (1978). The literature on cluster analysis. Multivariate behavioral research, 13(3), 271-295.

Bonne, K., Vermeir, I., Bergeaud-Blackler, F., &Verbeke, W. (2007). Determinants of halal meat consumption in France. British Food Journal, 109(5), 367-386.

Borsellino, V., Schimmenti, E., & El Bilali, H. (2020). Agri-food markets towards sustainable patterns. Sustainability, 12(6), 2193.

Burlingame, B., & Dernini, S. (2010). Sustainable diets and biodiversity. International Scientific Symposium. Rome. 2010

Carlsson-Kanyama, A., & González, A. D. (2009). Potential contributions of food consumption patterns to climate change. The American journal of clinical nutrition, 89(5), 1704S-1709S.

Cerroni, S., Nayga, R.M., Pappalardo, G., Yang, W., 2022. Malleability of food values amid the COVID-19 pandemic. Eur. Rev. Agric. Econ. 49, 472–498.

Chakraburty, S., Alizadeh, M., Handique, L., Altun, E., & Hamedani, G. G. (2021). A new extension of Odd Half-Cauchy Family of Distributions: properties and applications with regression modeling.

Chen, M. F. (2015). An examination of the value-belief-norm theory model in predicting pro-environmental behaviour in T aiwan. Asian Journal of Social Psychology, 18(2), 145-151.

Conner, M., & Sparks, P. (2005). Theory of planned behaviour and health behaviour. Predicting health behaviour, 2(1), 121-162.

Cook, A. J., Fairweather, J. R., & Campbell, H. R. (2000). New Zealand farmer and grower intentions to use genetic engineering technology and organic production methods..

Daviet, R., & Nave, G. (2024). The Value of Genetic Data in Predicting Preferences: A Study of Food Taste. Journal of Marketing Research, 00222437241244736.

De Boer, J., Hoogland, C. T., & Boersema, J. J. (2007). Towards more sustainable food choices: Value priorities and motivational orientations. Food Quality and Preference, 18(7), 985-996.

Dubreuil, S. (2018). Sustainability Action Plan.

El Bilali, H., Strassner, C., & Ben Hassen, T. (2021). Sustainable agri-food systems: environment, economy, society, and policy. Sustainability, 13(11), 6260.

Furst, T., Connors, M., Bisogni, C. A., Sobal, J., & Falk, L. W. (1996). Food choice: a conceptual model of the process. Appetite, 26(3), 247-266.

Garnett, T. (2013). Food sustainability: problems, perspectives and solutions. Proceedings of the Nutrition Society, 72(1), 29–39. doi:10.1017/S0029665112002947

George, J. F. (2002). Influences on the intent to make Internet purchases. Internet Research, 12(2), 165-180.

Hair Jr, J. F. (2006). Successful strategies for teaching multivariate statistics. In Proceedings of the 7th International Conference on (pp. 1-5).

Hair, J. F., Black, W. C., Babin, J. B., Anderson, R. E., & Tatham, R. L. (2010). Multivariate data analysis (6th ed.). Pearson/Prentice Hall

Harrison, D. A., MykytynJr, P. P., & Riemenschneider, C. K. (1997). Executive decisions about adoption of information technology in small business: Theory and empirical tests. Information systems research, 8(2), 171-195.

Hassan, L. M., Shiu, E., & Parry, S. (2016). Addressing the cross-country applicability of the theory of planned behaviour (TPB): A structured review of multi-country TPB studies. Journal of Consumer Behaviour, 15(1), 72-86.

Honkanen, P., Olsen, S. O., & Verplanken, B. (2005). Intention to consume seafood—the importance of habit. Appetite, 45(2), 161-168.

Huang, L., Yoshida, Y., Li, Y., Cheng, N., Xue, J., & Long, Y. (2024). Sustainable lifestyle: Quantification and determining factors analysis of household carbon footprints in Japan. Energy Policy, 186, 114016.

Huynh, M. (2018). Bringing the islands to schools: improving student attitudes towards statistics (Doctoral dissertation, RMIT University). https://core.ac.uk/download/pdf/159140480.pdf

Ip, C. Y., Zhuge, T., Chang, Y. S., Huang, T. H., & Chen, Y. L. (2022). Exploring the determinants of nascent social entrepreneurial behaviour. International Journal of Environmental Research and Public Health, 19(6), 3556.

Kaiser, H. (1974). An index of factorial simplicity. Psychometrika, 39, 31-6.

Kumar, P., Raj, A., & Kumar, V. A. (2024). Approach to Reduce Agricultural Waste via Sustainable Agricultural Practices. In Valorization of Biomass Wastes for Environmental Sustainability: Green Practices for the Rural Circular Economy (pp. 21-50). Cham: Springer Nature Switzerland.

Locher, J. L., Yoels, W. C., Maurer, D., & Van Ells, J. (2005). Comfort foods: an exploratory journey into the social and emotional significance of food. Food & Foodways, 13(4), 273-297.

Lu, Y., Jenkins, A., Ferrier, R. C., Bailey, M., Gordon, I. J., Song, S., ...& Zhang, Z. (2015). Addressing China's grand challenge of achieving food security while ensuring environmental sustainability. Science advances, 1(1), e1400039.

Lusk, J. L., & Briggeman, B. C. (2009). Food values. American journal of agricultural economics, 91(1), 184-196.

Lusk, J.L., 2011. External validity of the food values scale. Food Qual. Prefer. 22, 452–462.

Magnusson, M. K., Arvola, A., Hursti, U. K. K., Åberg, L., &Sjödén, P. O. (2001). Attitudes towards organic foods among Swedish consumers. British food journal, 103(3), 209-227.

Martins, J. M., Moguluwa, S. C., Lucas, J. L., Iloka, C. B., & Mata, M. N. (2023). Does perceived quality mediate the relationship between country of origin image and consumer buying behaviour in Nigerian hypermarkets?. PloS one, 18(8), e0285438.

Mathieson, K. (1991). Predicting user intentions: comparing the technology acceptance model with the theory of planned behavior. Information systems research, 2(3), 173-191..

Mercês, Z. D. C. D., Salvadori, N. M., Evangelista, S. M., Cochlar, T. B., Strasburg, V. J., da Silva, V. L., & Oliveira, V. R. D. (2024). Effectiveness, Challenges, and Environmental Impacts of New Food Strategies with Plant and Animal Protein Products. Foods, 13(20), 3217.

Mhlophe, B., & Zulu, M. (2022). Social context factors and Consumer Innovativeness as drivers of organic food Adoption amongst Millennials in South Africa (Doctoral dissertation).

Pappalardo, G., Lusk, J.L., 2016. The role of beliefs in purchasing process of functional foods. Food Qual. Prefer. 53, 151–158.

Pavlou, P. A. (2002, August). WHAT DRIVES ELECTRONIC COMMERCE? A THEORY OF PLANNED BEHAVIOR PERSPECTIVE. In Academy of management proceedings (Vol. 2002, No. 1, pp. A1-A6). Briarcliff Manor, NY 10510: Academy of Management.

Piracci, G., Casini, L., Contini, C., Stancu, C. M., & Lähteenmäki, L. (2023). Identifying key attributes in sustainable food choices: An analysis using the food values framework. Journal of Cleaner Production, 416, 137924.

Prasanna, S., Verma, P., & Bodh, S. (2024). The role of food industries in sustainability transition: a review. Environment, Development and Sustainability, 1-21.

Purvis, B., Mao, Y., & Robinson, D. (2019). Three pillars of sustainability: in search of conceptual origins. Sustainability science, 14, 681-695.

Ragaert, P., Verbeke, W., Devlieghere, F., & Debevere, J. (2004). Consumer perception and choice of minimally processed vegetables and packaged fruits. Food quality and preference, 15(3), 259-270.

Raghavendra, K. J., John, J., Jacob, D., Rajendran, T., Prusty, A. K., Ansari, M. A., ...& Dutta, D. (2024). Unraveling determinants of integrated farming systems adoption for sustainable livelihood and dietary diversity. Frontiers in Nutrition, 11, 1264658.

Raj, S., Choudhury, M., Mishra, B. B., & Samal, A. R. (2024). The role of extrinsic and intrinsic factors on the purchase intention toward processed food: a cross-sectional study in India. International Journal of Business and Emerging Markets, 16(1), 21-49.

Reisch, L., Eberle, U., & Lorek, S. (2013). Sustainable food consumption: an overview of contemporary issues and policies. Sustainability: Science, Practice and Policy, 9(2), 7-25.

Rhodes, R. E., & Yao, C. A. (2015). Models accounting for intention-behavior discordance in the physical activity domain: a user's guide, content overview, and review of current evidence. International Journal of Behavioral Nutrition and Physical Activity, 12, 1-14.

Rupp, H. (2003). Chemical and physical hazards produced during food processing, storage, and preparation (pp. 233-263). Food Safety Handbook.

Şimşek, E. K., Kara, M., Kalıpçı, M. B., & Eren, R. (2024). Sustainability and the Food Industry: A Bibliometric Analysis. Sustainability, 16(7), 3070.

Sparks, P., Shepherd, R., & Frewer, L. J. (1995). Assessing and structuring attitudes toward the use of gene technology in food production: The role of perceived ethical obligation. Basic and applied social psychology, 16(3), 267-285

Sparks, P., Shepherd, R., & Frewer, L. J. (1995). Assessing and structuring attitudes toward the use of gene technology in food production: The role of perceived ethical obligation. Basic and applied social psychology, 16(3), 267-285

Srinivasan, S., Galvez, A., Krieger, R., Sebo, A., Mckever, M., Nestico, D., ...& Everitt, T. (2024). Factors that facilitate consumer uptake of sustainable dietary patterns in Western countries: a scoping review. Frontiers in Sustainable Food Systems, 8, 1333742.

Suh, B., & Han, I. (2003). The impact of customer trust and perception of security control on the acceptance of electronic commerce. International Journal of electronic commerce, 7(3), 135-161.

Sun, S., Law, R., & Schuckert, M. (2020). Mediating effects of attitude, subjective norms and perceived behavioural control for mobile payment-based hotel reservations. International Journal of Hospitality Management, 84, 102331.

Tarkiainen, A., & Sundqvist, S. (2005). Subjective norms, attitudes and intentions of Finnish consumers in buying organic food. British food journal, 107(11), 808-822..

Tongco, M. D. C. (2007). Purposive sampling as a tool for informant selection.

Van Esch, P., Heller, J., & Northey, G. (2019). The effects of inner packaging color on the desirability of food. Journal of Retailing and Consumer Services, 50, 94-102.

Varzakas, T., & Smaoui, S. (2024). Global food security and sustainability issues: the road to 2030 from nutrition and sustainable healthy diets to food systems change. Foods, 13(2), 306.

Verain, M. C., Bartels, J., Dagevos, H., Sijtsema, S. J., Onwezen, M. C., & Antonides, G. (2012). Segments of sustainable food consumers: a literature review. International Journal of Consumer Studies, 36(2), 123-132.

Verbeke, W., &Vackier, I. (2005). Individual determinants of fish consumption: application of the theory of planned behaviour. Appetite, 44(1), 67-82.

Vermeir, I., & Verbeke, W. (2008). Sustainable food consumption among young adults in Belgium: Theory of planned behaviour and the role of confidence and values. Ecological economics, 64(3), 542-553.

Vermeir, I., Weijters, B., De Houwer, J., Geuens, M., Slabbinck, H., Spruyt, A., ...& Verbeke, W. (2020). Environmentally sustainable food consumption: A review and research agenda from a goal-directed perspective. Frontiers in psychology, 11, 1603.

Yang, Y., Hobbs, J.E., 2020. Food values and heterogeneous consumer responses to nanotechnology. Can. J. Agric. Econ. 68, 289–313.

14

Migrant Entrepreneurship: The Six-decade Long Journey of Venture Sustainability and Waste Reduction in India using Intellectual Capital and Firm Capability

Rajat Sharma[1], Muskan Singh[2], Neeraj Sharma[2], and Firas Rifai[3]

[1]Entrepreneurship Development Institute of India-Ahmedabad
[2]Graphic Era Deemed to be University, India
[3]Al-Zaytoonah University of Jordan, Jordan
E-mail: Sharmarajat126@gmail.com;
Muskansinghofficial1998@gmail.com;
neerajresearcher@gmail.com; frifai@gmail.com

Abstract

There are certain migrant entrepreneurs who have overcome all the difficulties and made their venture a big success. This research study explores the understudied topic of immigrant entrepreneurs and how these immigrants have created a viable business venture employing firm capabilities and intellectual capital that resulted in the venture enterprises' successful expansion over a period of six decades. The chapter uses an original, interpretative, case-based technique to convey a thorough understanding of the many difficulties experienced by migrant entrepreneurs and how they overcame them, and even curbed the waste generated. The chapter also highlights many transitions and changes that occurred in the six decades following India's partition by focusing on one of the rising industries in India, the watch business, and how the watch industry is using waste The chapter fills a gap in the literature about sustainable business ventures started by immigrant entrepreneurs in developing nations and how the entrepreneur overcame different challenges

DOI: 10.1201/9788743806905-14

and brought sustainability to the business and the ecosystem by not just efficiently using the resources but also reducing the waste that is generated through scrappage.

Keywords: Migrant entrepreneur, intellectual capital, waste, sustainability, watch industry.

14.1 Introduction

14.1.1 Migrant entrepreneurs

Migrant is defined as a person who resides in a country other than his home country for a period of at least one year (Castles & Miller, 2009). "Immigrant" is the term given to a person who was born abroad, while an immigrant's offspring is termed a "second generation" immigrant. The discussion on migrants has been one of the keen interest areas of scholars across various academic's disciplines but has been not explored in great depth in the field of entrepreneurship (Dana, 2007). Thus, migrant entrepreneurship has been acknowledged as an entrepreneurial venture undertaken by immigrants for their sustainability in the migrated country (Nazareno et al., 2018).

Migrant entrepreneurship has been a popular research concept in terms of developed economies such as UK (Patterson, 1963; Ward & Jenkins, 1984), USA (Bonacich, 1987; Light, 1972; Waldinger, 1984), and Canada (Dana, 1993). Although immigrant entrepreneurship literature is a large body of work, existing scholarly literature is highly fragmented and contextual. Many case studies are ungeneralizable (Dana, 2007) and primarily focus on migrant entrepreneurs who have migrated from developing economies to developed ones. For example, (Azmat & Fujimoto, 2016) studied Indian immigrants carrying out entrepreneurial ventures in Australia; (Pio et al., 2014) examined Indian immigrants practicing entrepreneurship in New Zealand; and Rahman & Lian (2011) explored immigrants of Bangladesh conducting ventures in the developed nation of Japan. Whereas this study is unique in its own way, it is, to the best of our knowledge, the first to examine migrant entrepreneurs in the context of two emerging economies – India and Pakistan – and to highlight the crucial role played by intellectual capital and firm capabilities in bringing economic sustainability to business ventures. The study is further supported by a case study of a marginalized immigrant entrepreneur Mr. Mehta, who migrated from what is now called Pakistan during the time when India got independence. The study maps his business journey in India over a period of six decades by a unique interpretative case-based method.

Furthermore, the objective of this research was to explore the motivations behind migrant entrepreneurship and investigate the extent to which they recognize opportunities, improve firm performance, and ensure business sustainability by the right application of intellectual and firm capabilities and surpass the hurdles during the process of venture creation and sustainability. Our inspiration for this research work is the suppressed level of migrant entrepreneurs especially in emerging economies due to negative socio culture and economic externalities (Nijkamp et al., 2010). Moreover, migrant people are viewed as incompetent than as a source of opportunity creators for the emerging economies (Sinkovics & Reuber, 2021), (Nijkamp et al., 2010). Thus, to cater to the problems of migrant entrepreneurs, the study provides an opportunity to advance our knowledge of migrant entrepreneurship in an emerging economy like India and sensitize people about migrant entrepreneurship and contribute to the field by answering the following research question:

RQ – What is the relationship between the reasons to engage in entrepreneurial activity and the subsequent role of intellectual capital and firm capabilities in sustaining and enhancing the firm performance of Indian immigrant entrepreneurs?

14.2 Literature Review

14.2.1 Migrant

Migrant people are a valuable source of new opportunities for countries, especially emerging ones, as they contribute to enhancing economic outcomes, boosting income, and reducing unemployment levels (Desiderio & Salt, 2010). Migrants also have a higher probability of initiating new ventures, further supporting the economy through their ventures (Mestres, 2010). Stone & Stubbs (2007) state that venture engagement provides better income generation opportunities to migrant people as compared to jobs. Thus, migrant entrepreneurship accelerates the growth of the emerging economies (Fatoki & Patswawairi, 2012). In majority of emerging economies, migrant people are engaged in entrepreneurial activities that are unrecognized and are considered low level (Rath, 2000), their growth is stagnant and their size is generally small, and their chances of growth are quite low due to different kinds of barriers (Desiderio, 2014). Moreover, in emerging economies, migrant people tend to initiate entrepreneurial ventures out of necessity and unacceptability from the society than their personal interest; therefore, it becomes crucial to explore and enhance the literature in context of migrant entrepreneurs in emerging economy context (Aliaga-Isla & Rialp, 2013).

14.2.2 Resource-based view (RBV)

Resource-based view (RBV) has been defined as a managerial framework used to determine the strategic resources a firm can exploit to achieve sustainable competitive advantage (Battisti et al., 2022). To achieve this competitive advantage, a firm should have Valuable, Rare, Inimitable, and Organized (VRIO) attributes that foster firm performance and make the firm sustainable (Miethlich & Oldenburg, 2019). Sustainability of a venture is a cumbersome process and to achieve this resource-based view has been considered significant in relating Intellectual Capital with Firm Performance (Khaustova et al., 2020). This sustainability can be achieved using the assets that a firm possesses, which are either tangible or intellectual (Ahmad et al., 2022). The asset possession of any firm depends on the financial capability of the firm, and in the context of the majority of migrant entrepreneurs, intellectual capital is preferred for performance enhancement and achieving firm sustainability (Grande et al., 2011).

14.2.3 Intellectual capital (IC)

Galbraith (1969) stated that intellectual capital inculcates more than intellect and includes pure intellect that includes an action that is intellectual (Melendez, 2017). IC is not simply a fixed intangible asset, but an accumulation of all intangible assets of the firm. These intangible assets should be used to develop products and services that offer value (Bataineh et al., 2022). Thus, it can be stated that intangible assets play a crucial role in firm growth (Dust, 2008) and those firms which have more intangible assets sustain longer (Asiaei et al., 2020). For achieving sustainability, migrant entrepreneurs must rely on intellectual capital – the intangible assets – as financial constraints don't allow them to depend heavily on tangible assets. Therefore, the core three determinants of intellectual capital – human capital (HC), relational capital (RC), and structural capital (SC) – play a crucial role in boosting performance and assisting migrant entrepreneurs in achieving sustainability in a competitive, dynamic environment (Demartini & Beretta, 2020).

14.2.3.1 Human capital (HC)

Human capital has been defined as one of the crucial aspects of IC and a significant role player for a firm as an asset (You et al., 2021). Its components, such as human knowledge, employee creativity, and the way people use their abilities for the firm, are also considered in intellectual capital, making human capital an ideal dimension for a firm (Anwar et al., 2018). Although the human capital involved with migrant entrepreneurs may lack the required skill, but recognizes the need of the customer to a great extent and tries to fulfill the upcoming and changing needs of the buyers (Kloosterman, 2010). Thus, to implement human capital effectively, it is significant for a firm, especially initiated by migrant entrepreneurs, to have capable human capital to foster the firm performance (Eresia-Eke & Okerue, 2020).

14.2.3.2 Relational capital (RC)

Relational capital is defined as the relation a firm share with its internal and external stakeholders, which include employees, customers, material providers, strategic alliance partners, and industry associations (Venugopalan et al., 2018). RC of a firm is identified as those assets of a firm that have been aligned in a sequence and that manage the relationship a firm maintains with the dynamic surroundings in which its business operates (Kovach et al., 2015). The most important area for relational capital is the rapport of the firm, which is built on the basis of the relationships that have existed between the firm and its consumers (Stam et al., 2021). Consumers place their trust in a firm's reputation, which boosts the appeal of the trade between the firm and the client (Duan et al., 2023). Thus, developing a positive reputation with consumers lowers the perceived risk of an exchange and enhances the chances of firm sustainability (Xing et al., 2020).

14.2.3.3 Structural capital (SC)

Structural capital is a component of a firm that survives the departure of HC (Bontis et al., 2015) It is the medium that keeps the firm moving at the right pace (Marr, 2005). Furthermore, SC not only establishes knowledge acquisition systems, but collects and integrates that knowledge too (Alquershi et al., 2021). Patents, culture, database, copyright, and process are all parts of structural capital (Hejazi et al., 2016). Structural capital includes all nonhuman storehouses of information in a firm such as equipment, databases, and processes; therefore, it can be said that SC belongs to the entire firm and can be recorded and redistributed (Khan et al., 2022). Research states that a firm that possess great sense of its structural capital leads to development of an environment that motivates the employees to try new ideas, give them the opportunity to learn from failures, and innovate more often, thus enhancing the chances of firm sustainability (Soriya et al., 2022).

14.2.4 Firm capability (FC)

Researchers describe firm capability as a firm's ability to execute various tasks of a firm and to achieve their specific goal (Ye et al., 2022), and boost firm performance thus giving the firm a competitive edge over its competitors (Bamgbade et al., 2022). Further, researchers have identified elements of capabilities which impact firm performance and leads to creativity (Le & Ikram, 2022). Strategic planning and vision alignment ultimately assist in designing a successful business model that guides firm performance toward achieving sustainability (Carayannis et al., 2015).

Dynamic sustainability was popularized by Teece et al. (2007) and after almost one and a half decades, he organized the diverse firm capabilities into three groups: (1) recognition and appraisal of an opportunity (sensing); (2) capital utilization to resolve potential and hence to gain value (seizing); and (3) continuing regeneration (reconfiguration) (Teece, 2011). Based on this classification, the view of dynamic capability (DC) has been conceptualized which has been further related with entrepreneurship orientation (EO) in SME context to achieve better firm performance (Hsu andWang, 2010). After a critical review, this research work has identified dynamic capability (DC) and entrepreneurship orientation (EO) as firm level capabilities.

14.2.4.1 Dynamic capability

Dynamic capability refers to the ability of an enterprise to adapt the dynamic business environment by adopting the right skills and integrating the resources and reconfiguration them according to the need of the dynamic environment (Teece et al., 1997). Further dynamic capability has its determinants such as sensing capabilities, integration capabilities, and reconfiguration capabilities that help in surviving the dynamic environment (Khan & Lew, 2018).

14.2.4.1.1 Sensing capability

The dynamic business environment is persistently changing and it becomes crucial for the business to sense the opportunity in the right time to make a significant impact in the market (Karneli, 2023). This dimension is further explored by transforming the available opportunities through innovation for which the firm has to sense and select the right opportunity in the right time (Teece et al., 2007).

14.2.4.1.2 Integration capability

Integration capability is the capability of the firm which assists the firm in maintaining a strong relationship with the consumer, and based on this relation, the firm is able to better identify the market demand, integrate the required changes in production and innovation, and fulfill the demand of the consumer (Teece et al., 2007). Moreover, it can be referred as the capability to deploy the right knowledge and to help reconfigure the firm resources for better output (Buzzao & Rizzi, 2021).

14.2.4.1.3 Reconfiguration capability

Reconfiguration capabilities are the abilities of a firm that assist in reconfiguring the resources and implementing them more effectively to enhance firm performance by improving both technical and market innovation. These capabilities help make the firm a persistent innovator by amending the firm resources and restructuring them according to the dynamic environment (Moghaddam & Weber, 2021).

14.2.4.1.4 Dynamic capability a crucial element for achieving SME sustainability

"Dynamic capability (DC) is seen as a firm's potential to systematically solve problems through more dynamic applications and adjustments of the firm's knowledge base, formed by knowledge sensing capacity, knowledge seizing capacity and knowledge reconfiguring capacity" (Teece et al., 2007; Li & Liu, 2014). Dynamic capability assists a firm in developing the knowledge-sensing capacity that results in identifying the valuable opportunities and safeguard from the threats present in the dynamic business environment (Khan et al., 2020). With the assistance of DC, firms are able to acquire sufficient knowledge and this knowledge further help in enhancing the seizing capacity of the firm which results in swift and wise decision-making with respect to investment into various opportunities (Hsu & Wang, 2012). These opportunities are further tapped by optimally reconfiguring the firm resources according to the constant changing needs (Ovuakporie et al., 2021). Furthermore, dynamic capabilities help SMEs to achieve sustainable performance in a cost-effective expenditure and by optimum utilization of resources possessed by the firm (Boons & Lüdeke-Freund, 2013; Daily & Huang, 2001).

Resources are scarce and the business environment is persistently evolving; therefore, majority of the firms have started implementing entrepreneurship orientation (EO). It was further found that EO has a crucial effect on operations of a firm and a culture that further assist in building strategy and boosting firm performance (Ribeiro et al., 2021).

14.2.5 Sustainability

The concept of sustainability has been associated with Wes Jackson's work on agriculture (Jackson, 2002) which was later on taken up by Brundtland Commission Report. The report defined sustainability in terms of development of the human ecology as "development that meets the needs of the present generation without compromising the ability of future generations to meet their own needs" (WECD, 1987). In terms of firms, sustainability is defined as the need fulfillment of the people who are directly or indirectly associated with the firm without compromising the future requirements of the associated people (Dyllick & Hockerts, 2002).

Sustainability is further classified into three types: social, environmental, and economic sustainability (Choi & Ng, 2011; Leaniz and Bosque, 2013). Out of the three sustainability dimensions, literature has repeatedly focused on the environmental dimension because of the impact of rapid consumption on the environmental degradation and eco-system resource constraints (Choi & Ng, 2011; Sheth, 2011). Therefore, to expand the field of sustainability, our study focuses on economic sustainability of business and that too in a migrant SME context. Economic dimension of sustainability refers to the organizational capability to create value and enhance financial performance (Leaniz and Bosque, 2013). Further, the previous research work has outlined two aspects of economic sustainability – (1) cost reductions and (2) economic interests of external stakeholders such as improvement in economic wellbeing and cost of living (Sheth, 2011). Leaniz and Bosque (2013) highlight that organizations involved in sharing of information with the stakeholders regarding their profits, achieve long-term success, and experience improved economic performance which further assist the firm to achieve economic sustainability.

14.2.5.1 SMEs and sustainability

SME is one of the growing fields in business research; still SME literature posits that irrespective of growing number of SMEs, the rate of SME failure is also increasing. Therefore, the sustainability of SMEs has become a key challenge to be addressed, especially when the SME is run by migrant people (Yeasmin, 2016). The concept of sustainability is still a new concept among Indian SMEs and many industries and businesses do not understand the benefits of sustainable practices. Therefore, it becomes crucial to associate migrant SMEs with firm performance and sustainability in an emerging economic context as SMEs have remained a valuable contributor in the GDP of emerging economies (Zubair & Brzozowski, 2018). Migrant SMEs are crucial contributors in the Indian context, as many people have migrated from neighboring countries to India after it gained independence (Webster & Haandrikman, 2017). Therefore, the upcoming section of the chapter highlights a crucial sector of the Indian economy in which a large number of SMEs run their ventures. Furthermore, the chapter is supported by case study evidence based on a venture run by a migrant entrepreneur, thus adding to the literature on migrant entrepreneurship.

14.2.6 Watch industry in India

When India gained independence in 1947, its manufacturing base was quite weak as a result of two centuries of economic exploitation by British rulers. As a result, the residents of India were persistently relying on goods manufactured outside India including the watches. Majority of these watches were manufactured in Switzerland and were in demand by the high net worth individuals residing in India. In India, the wrist watches were highly acknowledged but as they were expensive, the majority population of India was hesitant in purchasing them. This unsatisfied want of Indian citizens was widely acknowledged by the government in 1960s and after various rounds of brainstorming, the watch production was initiated in India with a motive of catering the masses and not just the classes. The India government gave a nod to an existing public-sector unit Hindustan Machine Tools (HMT) to initiate the manufacturing of watches. It gave a boost to the watch market in India and India witnessed a huge growth in the demand of watches especially wrist watches. The watches produced by the sole watch manufacturer, HMT, were affordable, trustworthy, and provided a sense of worthiness after wearing them. HMT made the watch culture prevalent in India with its diverse range in the market, which included models like Pilot and Sona, among which Janata was the most loved model by the Indian consumers due to its design and charm. Furthermore, HMT evolved over time and focused on mass production to meet the large demand of Indian consumers by efficiently using micro-sized components. However, the entry of Allwyn in 1981 and Titan in 1987 changed the fate of HMT. To worsen the situation, the LPG (liberalization, privatization, and globalization) reforms affected HMT, and foreign players entered the Indian watch market (Chattopadhyay & Bhawsar, 2017).

As significant players emerged in the watch market, intense competition ensued. Companies like Titan, due to its affordability, overtook HMT in the Indian market. Meanwhile, consumers with higher disposable incomes chose imported brands. HMT, unablFe to compete with both local and global brands, lost its prominence in the Indian market. Slowly, Titan took advantage of this and analyzed the Indian market carefully and offered a wide range of products at affordable prices leading to a shift in Indian consumer trust toward Titan. HMT, once the leader in the watch segment, became a laggard, while Titan ultimately emerged as the leader in India's watch market (Manikandan, 2021).

14.2.6.1 Recent trends in the Indian watch market

The Indian watch market has witnessed the exposure of watch brands from Japan, Switzerland, and Germany. The Japanese watch brands include Casio, Seiko, and Citizen. The popular Swiss brands dominating in India are Rado and Tissot; the famous German watch brands in India include Mont blanc and Wempe. These global brands are present in India because the watch industry of India is responsible for a production of more than 50 million watches per annum with figures crossing 65 billion dollars with a growth rate of 8–10% every year (Prabhakar, 2015). The Indian watch industry is now closely associated with personal wardrobes, and watch is now seen as an accessory (Kapferer & Laurent, 2016). This includes a wide range of classifications

Table 14.1 Different segments of watch industry in India

Segment A: watch with a price less than $15 fall in this segment and the labor force, students and wages-based earner prefer to buy these watches, the popular brand in this segment are FastTrack by Titan and Sonata by Tata (MP, 2022)	**Segment B**: Watches priced between $15–$150, this segment dominates in terms of sales, as a large chunk of the population in India belongs to middle and upper-middle class whose disposable income for watches falls within this category. Fossil, Police, Casio, and Titan dominate this segment (Donzé and Borel, 2019)
Segment C: Watch that is priced in the range $150– $1000, majority of rich people tend to lie in this parity. Although consumers in this segment prefer foreign brands but Titan is making its road in this segment too and is getting quite popular in its efforts. Citizen, Diesel dominate this sector (Donze, 2007).	**Segment D**: Watch that is priced above $1000 is considered in this segment and big businessman, celebrities are the people who purchase the watches in this segment and brands like Rado, Omega, and Rolex dominate this segment in terms of sales and buyer preference (Podaru, 2019).

Source: Author Compilation

such as formal watch, sports watches, chronograph-based watches, and fitness bands. In one way or another, the population of India is linked to watches. Depending on purchasing power and disposable income, the watch market segmentation is highlighted in Table 14.1.

14.2.6.2 Marketing a significant role player for the growth of watch industry

In the last decade (2010–2020), marketing has emerged as a key player for the success of Indian watch industry. There are more than fifty thousand outlets selling watches across India. But these outlets are now getting intense competition from online sellers such as Amazon, Flipkart, Myntra, and Ajio. In the last few years, the wristwatch segment of online retail has experienced a rapid growth. The main reason for this growth has been an enormous increase in the number of Internet users in the country, as Internet penetration has expanded across India, especially after 2015 when the current government launched the Digital India Scheme (a scheme that aims at making India connected through digital platforms). Competition in this sector is getting intense, and the majority of studies consider online platforms as facilitators of the watch business. Therefore, this study is unique and considers the journey of a migrant entrepreneur who has been in the watch business for more than six decades and is successfully running the venture despite global and local challenges.

14.2.6.3 Reducing waste

The watch industry is focusing on reducing waste by using the worn-out parts of the existing watches that are generated through a circular economy practice of reducing, reuse, and recycling. Furthermore, by using these waste parts, the watch industry is saving 10% of the costs they initially spent. Thus, efficient waste utilization has become the new trend in the watch industry, achieved through the effective use of human capital that uses the collected waste parts in the most precise manner. This not only allows the crafting of watches from these components but also helps fulfill the dreams of people living below the poverty line by providing them with watches at an affordable price. The major parts of these watches are made from waste components of the existing watches, which would otherwise be pure scrap lying in the factory.

14.3 Methodology

In this study, we have adapted to an interpretative case-based method which is composed of different rounds of questioning and observation session with the target audience in order to generate a rich and deep understanding about the research question. In this case-based approach, the researcher has interviewed Mr. Mehta the protagonist of the case who is also a migrant entrepreneur belonging to the migrant family from Pakistan now located in Dehradun, the capital of one of the newly formed state, Uttarakhand in India, and deep insights were generated on how Mehta's belonging to Punjabi community migrants builds and sustained their entrepreneurial venture over the period of six decades.

14.3.1 Case

Sukhdarshan Mehta, the person behind the establishment of Mehta Watch Company that was started in 1954 and is standing tall against time. The venture had a very humble beginning. Mr. Sukhdarshan was a migrant from Peshawar, Pakistan, and settled in Dehradun, the capital of Uttarakhand, a state in India. Mr. Sukhdarshan wanted to do something to sustain his family so he purchased a small shop near the clock tower of Dehradun city and started learning and practicing watch repairing. As Sukhdarshan belong to a Punjabi community, he was not well accepted by other businessmen of the town who belong to the (Baniya) business community of the town. Irrespective of this, Sukhdarshan believed in his hard work and kept working. This migrant entrepreneur was so indulged in his work and was so inclined toward his customer service that sometimes he returns home post-midnight. Sukhdarshan had six children and the eldest one among them was Mr. Harish Mehta, who is the protagonist of this case. Mr. Harish always supported his father in his school time and even handled business at the age of 22 when Mr. Sukhdarshan all of a sudden passed away in 1984. From that day, Harish Mehta never look back and kept the learnings of his father in his mind and served the customer with loyalty and a small shop that was purchased in 1954 grew to a small business empire in 2023. Mehta Watch Company not just sell and repair watches, rather they have two stores at the center of the Dehradun city. The store houses almost all the affordable and few luxury brands and also has more than 10 service centers of different mobile brands. The case of Mehta Watch Company is further addressed by an interpretative case-based approach and through several rounds of questions addresses the struggles of a migrant entrepreneur and how a venture is made sustainable for more than six decades. The watch industry is focusing on reducing waste by using the worn-out parts of the existing watches that are generated through a circular economy practice of reduce, reuse, and recycle. Furthermore, by

Table 14.2 Financial success of Mehta Watch Company

Year	Increment	Sales
1954–1964	0	100
1964–1974	10	100+10
1974–1984	12	100+12
1984–1994	18	100+18
1994–2004	25	100+25
2004–2014	30	100+30
2014–2022	32	100+32

Source: Author compilation Note: 100 has been considered as the base sales for Table 14.2

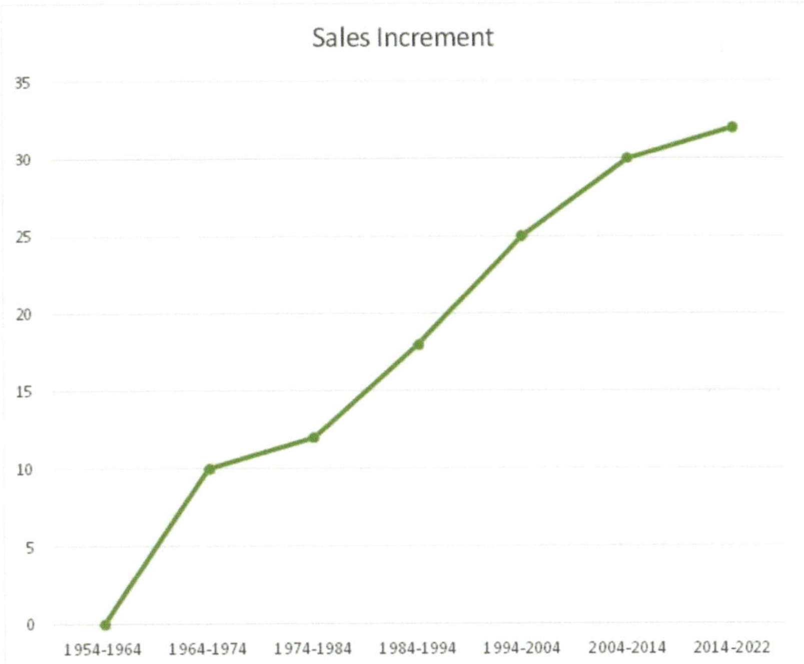

Figure 14.1 Sales increment of Mehta Watch Company [Author's compilation].

using these waste parts, the watch industry is saving 10% of the costs that were initially spent. Thus, efficient waste utilization is the new trend of watch industry, primarily driven by the effective utilization of human capital, which processes the collected waste parts in the most precise manner. This not only allows for the crafting watches from thsese components but also helps fulfill the dreams of people living below the poverty line by providing them with watches at affordable prices. The major parts of these watches are made from the waste components of existing watches, which would otherwise be pure scrap lying in the factory. As noted earlier, we also use waste in similar ways, as the worn-outs part at Mehta Watch Company are quite large in numbers. Therefore, we strive to make use of every waste part and provide repair facilities to even those customers who might not otherwise afford repairs. Thus, we use these waste parts and for the time being repair their watches free of cost as a sweet gesture through the utilization of waste parts. Table 14.2 and graph 1 reflect the financial success of Mehta Watch Company over six decades:

14.4 Conclusion

This chapter and case study highlight how the watch industry is using intellectual capital that is a combination of human capital, relational capital, and structural capital. Furthermore, the research discusses how all elements of intellectual capital contribute to sustaining the venture over time, even imparting longevity to the business venture. The research study also highlights how waste is being reused by the watch industry as a whole, including small family businesses involved in watch repairs. These businesses not only repair old watches but also reduce waste, thereby contributing to the sustainability goals of the entire industry.

References

Ahmad, M., Wu, Q., & Khattak, M. S. (2022). Intellectual capital, corporate social responsibility and sustainable competitive performance of small and medium-sized enterprises: mediating effects of organizational innovation. Kybernetes, (ahead-of-print).

Aliaga-Isla, R., & Rialp, A. (2013). Systematic review of immigrant entrepreneurship literature: previous findings and ways forward. Entrepreneurship & Regional Development, 25(9-10), 819- 844.

Anwar, M., Khan, S. Z., & Khan, N. U. (2018). Intellectual capital, entrepreneurial strategy and new ventures performance: Mediating role of competitive advantage. Business and Economic Review, 10(1), 63-93.

Asiaei, K., Barani, O., Bontis, N., & Arabahmadi, M. (2020). Unpacking the black box: how intrapreneurship intervenes in the intellectual capital-performance relationship?. Journal of Intellectual Capital, 21(6), 809-834.

Azmat, F., & Fujimoto, Y. (2016). Family embeddedness and entrepreneurship experience: A study of Indian migrant women entrepreneurs in Australia. Entrepreneurship & Regional Development, 28(9-10), 630-656.

Bamgbade, J. A., Nawi, M. N. M., Kamaruddeen, A. M., Adeleke, A. Q., & Salimon, M. G. (2022). Building sustainability in the construction industry through firm capabilities, technology and business innovativeness: empirical evidence from Malaysia. International journal of construction management, 22(3), 473-488.

Bartolacci, F., Caputo, A., & Soverchia, M. (2020). Sustainability and financial performance of small and medium sized enterprises: A bibliometric and systematic literature review. Business Strategy and the Environment, 29(3), 1297-1309.

Bataineh, H., Abbadi, S. S., Alabood, E., & Alkurdi, A. (2022). The effect of intellectual capital on firm performance: the mediating role of family management. Journal of Islamic Accounting and Business Research, 13(5), 845-863.

Battisti, E., Nirino, N., Leonidou, E., & Thrassou, A. (2022). Corporate venture capital and CSR performance: An extended resource based view's perspective. Journal of Business

Bonacich, E. (1987). "Making it" in America: A social evaluation of the ethics of immigrant entrepreneurship. Sociological Perspectives, 30(4), 446-466.

Bontis, N., Janošević, S., & Dženopoljac, V. (2015). Intellectual capital in Serbia's hotel industry. International Journal of Contemporary Hospitality Management, 27(6), 1365-1384.

Boons, F., & Lüdeke-Freund, F. (2013). Business models for sustainable innovation: state-of-the- art and steps towards a research agenda. Journal of Cleaner production, 45, 9-19.

Buzzao G., & Rizzi, F. (2021). On the conceptualization and measurement of dynamic capabilities for sustainability: Building theory through a systematic literature review. Business Strategy and the Environment, 30(1), 135-175.

Carayannis, E. G., Sindakis, S., & Walter, C. (2015). Business model innovation as lever of organizational sustainability. The Journal of Technology Transfer, 40, 85-104.

Castles, S., & Miller, M. J. (2009). Migration in the Asia-Pacific region. Migration Information Source, 10.

Chattopadhyay, U., & Bhawsar, P. (2017). Effects of changing business environment on organization performance: The case of HMT Watches Ltd. South Asian Journal of Business and Management Cases, 6(1), 36-46.

Choi, S., & Ng, A. (2011). Environmental and economic dimensions of sustainability and price effects on consumer responses. Journal of business ethics, 104, 269-282.

Daily, B. F., & Huang, S. C. (2001). Achieving sustainability through attention to human resource factors in environmental management. International Journal of operations & production management, 21(12), 1539-1552.

Dana, L. P. (1993). An inquiry into culture and entrepreneurship: case studies of business creation among immigrants in Montreal. Journal of Small Business & Entrepreneurship, 10(4), 16-31.

Dana, L. P. (Ed.). (2007). Handbook of research on ethnic minority entrepreneurship: A co- evolutionary view on resource management. Edward Elgar Publishing.

Demartini, M. C., & Beretta, V. (2020). Intellectual capital and SMEs' performance: A structured literature review. Journal of Small Business Management, 58(2), 288-332.

Desiderio, M. V., & Salt, J. (2010). Main findings of the conference on entrepreneurship and employment creation of immigrants in OECD Countries, 9-10 June 2010, Paris. Open for Business: Migrant Entrepreneurship in OECD Countries.

Desiderio, M. V., & Weinar, A. (2014). Supporting immigrant integration in Europe?: Developing the governance for diaspora engagement.

Donze, P. Y. (2007). Joint venture companies and technology transfer from the West to Japan: Citizen Watch Co and the Swiss watchmakers (1890–1940).)∧(eds), Book Joint venture companies and technology transfer from the West to Japan: Citizen Watch Co and the Swiss watchmakers (1890–1940).

Duan, Y., Chen, Y., Liu, S., Wong, C. S., Yang, M., & Mu, C. (2023). The moderating effect of leadership empowerment on relational capital and firms' innovation performance in the

Dyllick, T., & Hockerts, K. (2002). Beyond the business case for corporate sustainability. Business strategy and the environment, 11(2), 130-141.

entrepreneurial ecosystem: Evidence from China. Journal of Intellectual Capital, 24(1), 306-336.

Durst, S. (2008). The relevance of intangible assets in German SMEs. Journal of Intellectual Capital, 9(3), 410-432.

Eresia-Eke, C., & Okerue, C. (2020). Owner-manager human capital and business coping ability in African immigrant small businesses in Southern Africa. Development Southern Africa, 37(2), 348-361.

Fatoki, O., & Patswawairi, T. (2012). The motivations and obstacles to immigrant entrepreneurship in South Africa. Journal of Social Sciences, 32(2), 133-142.

Gates, S., & Langevin, P. (2010). Human capital measures, strategy, and performance: HR managers' perceptions. Accounting, Auditing & Accountability Journal, 23(1), 111-132.

Grande, J., Madsen, E. L., & Borch, O. J. (2011). The relationship between resources, entrepreneurial orientation and performance in farm-based ventures. Entrepreneurship and Regional Development, 23(3-4), 89-111.

Hejazi, R., Ghanbari, M., & Alipour, M. (2016). Intellectual, human and structural capital effects on firm performance as measured by Tobin's Q. Knowledge and Process Management, 23(4), 259-273.

Hsu, L. C., & Wang, C. H. (2012). Clarifying the effect of intellectual capital on performance: the mediating role of dynamic capability. British journal of management, 23(2), 179-205.

Jackson, W. (2002). Natural systems agriculture: a truly radical alternative. Agriculture, ecosystems & environment, 88(2), 111-117.

Kapferer, J. N., & Laurent, G. (2016). Where do consumers think luxury begins? A study of perceived minimum price for 21 luxury goods in 7 countries. Journal of Business

Karneli, O. (2023). The Role of Adhocratic Leadership in Facing the Changing Business Environment. Journal of Contemporary Administration and Management (ADMAN), 1(2), 77-83.

Khan, A. M., Arafat, M. Y., & Raushan, M. A. (2022). Analysing the entrepreneurial intentions through intellectual capital: evidences from India. International Journal of Intelligent

Khan, O., Daddi, T., & Iraldo, F. (2020). Microfoundations of dynamic capabilities: Insights from a circular economy business cases. Business Strategy and the Environment, 29(3), 1479- 1493.

Khan, Z., & Lew, Y. K. (2018). Post-entry survival of developing economy international new ventures: A dynamic capability perspective. International Business Review, 27(1), 149-160.

Khaustova, Y., Durmanov, A., Dubinina, M., Yurchenko, O., & Cherkesova, E. (2020). Quality of strategic business management in the aspect of growing the role of intellectual

Kloosterman, R. C. (2010). Matching opportunities with resources: A framework for analysing (migrant) entrepreneurship from a mixed embeddedness perspective. Entrepreneurship and Regional Development, 22(1), 25-45.

Kovach, J. J., Hora, M., Manikas, A., & Patel, P. C. (2015). Firm performance in dynamic environments: The role of operational slack and operational scope. Journal of Operations Management, 37, 1-12.

Le, T. T., & Ikram, M. (2022). Do sustainability innovation and firm competitiveness help improve firm performance? Evidence from the SME sector in vietnam. Sustainable Production and Consumption, 29, 588-599.

Leon, P. (2001). Four pillars of financial sustainability. Nature Conservancy.

Li, D. Y., & Liu, J. (2014). Dynamic capabilities, environmental dynamism, and competitive advantage: Evidence from China. Journal of business research, 67(1), 2793-2799.

Li, X., Yang, J., Liu, H., & Zhuang, X. (2021). Entrepreneurial orientation and green management in an emerging economy: The moderating effects of social legitimacy and ownership type. Journal of Cleaner Production, 316, 128293.

Light, I. (1972). Ethnic enterprise in America. Berkeley, Los Angeles, London: University of California Press.

Manikandan, K. S. (2021). Titan Company Limited: growth strategy. Emerald Emerging Markets Case Studies, 11(3), 1-38.

Marr, B., & Moustaghfir, K. (2005). Defining intellectual capital: a three-dimensional approach. Management decision, 43(9), 1114-1128.

Melendez, D. (2017). The impact of intellectual capital on firm's performance (Bachelor's thesis, D. Melendez).

Mestres, J. (2010). Migrant entrepreneurship in OECD countries and its contribution to employment.

Miethlich, B., & Oldenburg, A. G. (2019). How social inclusion promotes sales: An analysis of the example of employing people with disabilities. Journal of Marketing Research and Case Studies, 2019, 1-15.

Moghaddam, K., & Weber, T. (2021). Immigrants and international entrepreneurship: Transnational dynamic capability and dual social network embeddedness. In Empirical international entrepreneurship: A handbook of methods, approaches, and applications (pp. 9- 25). Cham: Springer International Publishing.

Nazareno, J., Zhou, M., & You, T. (2018). Global dynamics of immigrant entrepreneurship: Changing trends, ethnonational variations, and reconceptualizations. International Journal of Entrepreneurial Behavior & Research, 25(5), 780-800.

Nijkamp, P., Sahin, M., & Baycan-Levent, T. (2010). Migrant entrepreneurship and new urban economic opportunities: identification of critical success factors by means of qualitative pattern recognition analysis. Tijdschrift voor economische en sociale geografie, 101(4), 371-391.

Ovuakporie, O. D., Pillai, K. G., Wang, C., & Wei, Y. (2021). Differential moderating effects of strategic and operational reconfiguration on the relationship between open innovation practices and innovation performance. Research Policy, 50(1), 104146.

Patterson, S. (1963). Dark strangers: a sociological study of the absorption of a recent West Indian migrant group in Brixton, South London. (No Title).

Pio, E., & Essers, C. (2014). Professional migrant women decentring otherness: A transnational perspective. British Journal of Management, 25(2), 252-265.

Podaru, D. N. (2019). THE IMPORTANCE OF ICONICITY IN THE LUXURY WATCHES INDUSTRY. LIMBA ŞI LITERATURA–REPERE IDENTITARE ÎN CONTEXT EUROPEAN, (24), 202-209.

Prabhakar, A. (2015). Growth in the Indian watch industry. Images Retail, 172-176.

Rahman, M. M., & Lian, K. F. (2011). The development of migrant entrepreneurship in Japan: Case of Bangladeshis. Journal of International Migration and Integration, 12, 253-274.

Rath, J. (Ed.). (2000). Immigrant businesses (pp. 1-19). London: Palgrave Macmillan.

Ribeiro, M. A., Adam, I., Kimbu, A. N., Afenyo-Agbe, E., Adeola, O., Figueroa-Domecq, C., & de Jong, A. (2021). Women entrepreneurship orientation, networks and firm performance in the tourism industry in resource-scarce contexts. Tourism Management, 86, 104343.

Sheth, J. N. (2011). Impact of emerging markets on marketing: Rethinking existing perspectives and practices. Journal of marketing, 75(4), 166-182.

Sinkovics, N., & Reuber, A. R. (2021). Beyond disciplinary silos: A systematic analysis of the migrant entrepreneurship literature. Journal of World Business, 56(4), 101223.

Soriya, S., Mittal, R., & Lochab, A. (2022). Measuring role of IC in SMEs' profitability: evidence from India. International Journal of Learning and Intellectual Capital, 19(6), 584-607.

Stam, E., & Van de Ven, A. (2021). Entrepreneurial ecosystem elements. Small business economics, 56, 809-832.

Stone, I., & Stubbs, C. (2007). Enterprising expatriates: lifestyle migration and entrepreneurship in rural southern Europe. Entrepreneurship and regional development, 19(5), 433-450.

Teece, D. J. (2007). Explicating dynamic capabilities: the nature and microfoundations of (sustainable) enterprise performance. Strategic management journal, 28(13), 1319-1350.

D. J. (2011). Achieving integration of the business school curriculum using the dynamic capabilities framework. Journal of Management Development, 30(5), 499-518.

Teece, D. J., Pisano, G., & Shuen, A. (1997). Dynamic capabilities and strategic management. Strategic management journal, 18(7), 509-533.

Venugopalan, M., Sisodia, G. S., & Rajeevkumar, P. (2018). Business relational capital and firm performance: an insight from Indian textile industry. International Journal of Learning and Intellectual Capital, 15(4), 341-362..

Waldinger, R. (1984). Immigrant enterprise in the New York garment industry. Social Problems, 32(1), 60-71.

Ward, R., & Jenkins, R. (Eds.). (1984). Ethnic communities in business: Strategies for economic survival. Cambridge University Press.

Webster, N. A., & Haandrikman, K. (2017, January). Thai women entrepreneurs in Sweden: Critical perspectives on migrant small businesses. In Women's Studies International Forum (Vol. 60, pp. 17-27). Pergamon.

World Commission on Environment, Development. Advisory Panel on Food Security, & World Commission on Environment. (1987). Food 2000: Global Policies for Sustainable Agriculture: a Report of the Advisory Panel on Food Security, Agriculture, Forestry, and Environment to the World Commission on Environment and Development. London; New Jersey: Zed Books.

Xing, Y., Liu, Y., Boojihawon, D. K., & Tarba, S. (2020). Entrepreneurial team and strategic agility: A conceptual framework and research agenda. Human Resource Management Review, 30(1), 100696.

Yeasmin, N. (2016). The determinants of sustainable entrepreneurship of immigrants in Lapland: An analysis of theoretical factors. Entrepreneurial Business and Economics Review, 4(1), 129- 159.

You, S., Zhou, K. Z., & Jia, L. (2021). How does human capital foster product innovation? The contingent roles of industry cluster features. Journal of Business Research, 130, 335-347.

Zubair, M., & Brzozowski, J. (2018). Entrepreneurs from recent migrant communities and their business sustainability. Sociologica, 12(2), 57-72.

15

Machine Learning Base Model for Waste Management

Anuram Krishna E. S.[1], Preethi Nanjundan[2], and Lijo Thomas[3]

[1]Department of Global Financé Analytics, Christ University, India
[2]Department of Data Science, Christ University, India
[3]Department of Psychology, Christ University, India
E-mail: anuram.es@msgfa.christuniversity.in; preethi.n@christuniversity.in; lijo.thomas@christuniversity.in

Abstract

Waste management is an important global challenge, where increasing urbanization and industrialization lead to high waste production. Traditional waste landfill methods, including manual sorting and fixed collection plans often result in disabilities, increased costs, and environmental decline. Integration of machine learning (ML) and artificial intelligence (AI) in waste management systems provides a transformative approach to solve these problems. This chapter examines the role of ML in automatic waste sorting, future indication analysis, collection passport optimization, and converting waste-to-energy. Case studies from Singapore, San Francisco, and large waste management companies highlight ML's real-world applications to create smart and more sustainable waste management systems. Singapore's AI-operated system and San Francisco's policy-based waste management models show ML's ability to adapt to comparative analysis resources and reduce landfill addiction. The findings show that AI-controlled waste management leads to high recycling speeds, reduces greenhouse gas emissions, and more efficient resource allocation. Despite challenges such as high implementation costs, regulatory concerns, and privacy problems, the future of waste management lies in AI-operated automation, blockchain integration for waste tracking, and AI-cleaned waste solutions.

DOI: 10.1201/9788743806905-15

Keywords: Waste Managment, Artificial Intelligence, Sustainable Waste Management

15.1 Introduction

15.1.1 Required waste crisis and smart solution required

Waste management has become a pressing global issue due to rapid urbanization, industrial expansion, and population growth. The increase in waste production has placed considerable pressure on municipalities, industries, and environmental agencies to find efficient and sustainable solutions. According to the World Bank's *What A Waste 2.0* report, the world generated 2.01 billion tons of municipal stable waste (MSW) in 2016, and the number is estimated to increase to 3.4 billion tons by 2050. Improper handling of waste has led to environmental pollution and severe health risks.

Despite the progress of waste management techniques, only 67% of the global waste is properly controlled, while the remaining 33% are either untreated or dumped illegally. This error management contributes to serious environmental issues such as soil pollution, water pollution, and greenhouse gas emissions. In many developing countries, the waste management systems struggle with chronic infrastructure issues, a lack of technical integration, and insufficient regulatory enforcement. As a result, there is an urgent need for innovative, data–driven, and automated waste management systems that can improve efficiency and sustainability

15.1.2 Challenges in traditional waste management

Traditional waste management relies heavily on manual labor, outdated collection methods, and static disposal programs. One of the biggest challenges municipalities and industries face is incompetent waste collection. Many waste collection systems operate on a fixed plan instead of demand-based collections, leading to overflowing bins in some areas while others remain empty, causing unnecessary trips and excessive fuel consumption. This disability results in high operating costs and environmental burdens.

Another major problem is inappropriate waste sorting and contamination recycling. Many recycling programs suffer from poor waste insulation at the source, causing high pollution rates in recycling functions. In addition, illegal dumping is still a frequent problem in many areas, with irregular settlement contributions to air, water, and land pollution.

High operating costs further increase the problem. Waste management services require sufficient investments in collection, transport, processing, and landfill. Without data-driven insights, the municipalities are struggling to distribute resources efficiently, which leads to unnecessary expenses. In addition, the lack of real-time monitoring and data analysis by waste management managers prevents informed decisions to adapt waste management processes.

15.1.3 The role of machine learning (ML) in modern waste management

Machine learning (ML) has emerged as a transformative technique in waste management, offers solutions that automatically adjust to operations and reduce costs. ML algorithm can analyze huge amounts of sensors, smart trainers, GPS tracking systems, and satellite images, which can enable the waste management authorities to make informed decisions in real time.

One of the main benefits of ML in waste management is automatic waste classification and sorting. The AI-managed image recognition system can identify and classify various types of waste, improve recycling efficiency, and reduce human errors. It is especially useful in material recycling functions (MRFs), where AI-operated robotic weapons sort the waste based on physical structure. Studies have shown that the AI-based sorting system can achieve 99.95% accuracy, which can significantly increase the efficiency of recycling processes.

Predictive analysis is another important area where ML makes a difference. By analyzing historical waste production data, the ML models can predict future waste production trends. This helps cities and industries plan waste collection programs, optimize landfill use, and implement active waste deficiency strategies. For example, AI models such as LSTM (long short-term memory) networks and ARIMA (autoregressive integrated moving average) have been used successfully to predict waste production patterns with high accuracy, leading to better resource allocation. Traditional waste charcoal

15.1.4 Benefits of AI and ML in waste management

Applying AI and ML in waste management provides many benefits that solve large industries. Increased efficiency: AI-driven waste collection and sorting system significantly reduce human interference, which improves rapid treatment time and better resource allocation. Increased stability: ML enables better waste classification and high recycling speeds, supporting global stability efforts and a circular economy models. Cost savings: By adapting waste collection routes and reducing fuel consumption, the AI-driven systems help municipalities reduce operating costs by 20–30%. Automation and accuracy: Data-based AI models acquire high accuracy in waste classification, improve recycling efficiency, and reduce pollution. Reduction of environmental impact: AI-operated analysis supports better landfill management, waste-to-energy initiative, and emission tracking, which reduces ecological damage.

15.1.5 Machine learning objectives basic model for waste management

The purpose of this chapter is to find out different methods where ML changes waste management. It will provide insight into different ML models used for sorting, prediction, and adaptation.

Towards the end of this chapter, readers will gain a widespread understanding of how ML redefines the waste management strategies, making them more efficient, cost-effective, and environmentally durable.

15.2 Literature Review: ML-based Waste Management Model

In recent years, machine learning (ML) and artificial intelligence (AI) have brought a revolution in waste management by enabling automation, predictive analysis, recycling, and disposal processes. Traditional approaches to waste management are struggling with waste sorting, irregular collection programs, and disabilities in environmental pollution. The integration of the AI-operated model has provided a data-centered approach, leading to a more durable, cost-effective, and scalable waste management solution. This segment undergoes existing literature on ML applications in waste management, focusing on AI-based waste classification, future waste production, route adjustment, illegal dumping detection, and converting of waste into energy. The discussion includes discoveries from paper uploaded by the user, as well as insights from other research studies in the domain.

15.2.1 AI-driven waste management in smart cities

Chapter "Artificial Intelligence for Waste Management in Smart Cities: A Review" (user-uploaded documents) provides a comprehensive discussion about the role of AI and IoT-Saksham Waste Management Solutions. This emphasizes how computer vision-based deep learning models improve waste classification, how reinforcement learning optimizes waste collection routes, and how future analyses predict trends of waste. The paper emphasizes AI's role in creating smart cities where waste management is automatic and efficient, reducing operating costs and promoting stability.

One of the most important contributions in this study is the focus on AI interest in real-time surveillance systems. This discusses that AI algorithm-equipped drones, CCTV monitoring, and smart trainers can track illegal dumping activities, which can reduce environmental damage. This review creates a solid basis for further research when machine learning and big data analytics are integrated into waste management. Gupta et al. (2021) discovered AI-operated waste-sorting technologies in urban areas. Research emphasized the importance of high-quality training datasets to improve the accuracy of the AI model under real-world conditions.

15.2.2 AI-based waste classification and sorting

One of the most important advances in waste management is automatic waste sorting when using a machine learning algorithm. Traditional sorting methods are often dependent on manual labor, which is disabled, wrongly exposed, and expensive. Several studies have shown that the AI-based classification system improves traditional sorting techniques at speed and accuracy. Study by Thung & Yang (2020) developed Trashnet, to train deep learning models to classify the waste into categories such as paper, plastic, glass, and metal for an open-source dataset. When using the recreation and VGGNet architecture, the study reported classification accuracy of 92.3%. This dataset has been widely used in AI-based waste-sorting applications since then.

Similarly, Ramalingam et al. (2022) conducted a comparative analysis of various deep teaching models, and concluded that Yolo (you only see once) and quickly were most effective in real-time waste identity in the R-CNN smart recycling room. His study emphasized the role of high-resolution image sensors and cloud-based AI models to improve waste classification efficiency.

In addition, Kumar et al. (2021) suggested a hybrid deep learning model to combine CNN and supporting vocal machines (SVMs), which improved waste recognition rates in automated recycling systems. Their findings suggest that integration of multiple ML techniques can significantly increase the strength of classification models. Exact predictions on waste production can help municipalities optimize the waste collection program, allocate resources efficiently, and reduce garbage dump. The AI-operated future models analyze historical data to identify patterns in waste production. Landmark survey of Cortis et al. (2019) implemented long short-term memory (LSTM) network and AR.

In another study, Shah et al. (2021) analyzed smart bin sensor data from urban areas and trained the random forest and XGBoost models to estimate when the bins would be full. His findings indicated that the AI-driven prediction reduced unnecessary waste collections by 40%, leading to cost savings and environmental benefits. (2023) discovered how emotional analysis of social media can increase the waste spread model. Hans AI system discussed the discussion on public complaints and waste accumulation and predicted areas at high risk of waste overflow. This innovative approach showed how NLP (natural language processing) models can complement structured waste data for better forecasts.

15.2.3 AI for collecting smart waste and route optimization

Disabled waste collection routes resulted in fuel consumption, high operating costs, and unnecessary emissions. The AI models can adapt the routes to the dynamic waste plan and real-time data. Study of Ding et al. (2020) Use of reinforcement learning (DEEP Q -Network - DQN) to adapt waste collection car routes. Their model reduced the use of fuel by 26% compared to traditional static route approaches and increased collection efficiency by 32%. (2022) developed a genetic algorithm-based AI model, which analyzed traffic overloads, filling levels and weather conditions to adapt the collection roads. In a related study, Lee et al. (2021) discovered AI-operated fleet for waste collection control systems. Their research found that integration of AI with IoT–enabled competition competent smart rooms helped cities reduce 15–20% operating waste management costs annually.

15.2.4 AI for illegal dumping detection and environmental monitoring

Illegal dumping is a serious environmental problem, causing soil, water, and air pollution. The model for detecting AI-based monitoring and deviations is quickly adopted to deal with the problem. According to the study by Olivier et al. (2021), a system gained an accuracy of 98.3% in identifying illegal waste sites. Similarly, Chaudhary et al. (2022) deployed an AI-enhanced CCTV monitoring in Indian smart cities, which enabled real-time identification of the violations that determined waste outside the specified areas.

This approach improved compliance with waste rules. Another innovative study by Santos et al. (2023) used satellite images and deep learning models to track the cost of plastic pollution. Their AI-operated surveillance system provided valuable insight to decision makers in marine waste accumulation trends, providing targeted cleaning efforts.

Reviewed study depicts the transformation role of AI and ML in waste management, focusing on detection of automated waste sorting, future indication analysis, smart collection, and illegal dumping. Research has continuously shown that AI-operated systems increase efficiency, reduce costs, and support environmental stability. Nevertheless, despite these advances, data availability, high implementation costs, and challenges that regulatory obstacles face challenges. Future research should detect integration of blockchain technology for federal learning for waste traceability, AI-controlled waste-to-energy solutions, and analysis of privacy preservation.

15.3 The Role of Machine Learning in Waste Management

15.3.1 Introduction

The growing global population and rapid urbanization have led to a significant increase in waste production, causing serious environmental and logical challenges for municipalities and industries. The traditional waste management system, which depends mainly on manual work and static collection programs, has not been able to address the increasing waste crisis. Lack of insufficient waste insulation, disabled collection roads, and real-time monitoring contribute to excessive garbage dump, pollution, and high operating costs. Matchin learning (ML) has emerged as a powerful tool to bring revolution in waste management, provides automatic, data–driven, and scalable solutions. By detecting computer vision, predictive analytics, reinforcement learning, and deviations, ML-based models have increased waste classification, collection adjustment, and recycling processes. This progress not only improves efficiency but also promotes stability by supporting the principles of a circular economy. Figure 15.1 Indicating AI- based waste management process

This segment examines the role of waste management, focusing on automatic waste sorting, future waste production model, AI-driven waste collection, and environmental monitoring. The discussion highlights the challenges in different ML techniques, their real-world applications, benefits, and implementation.

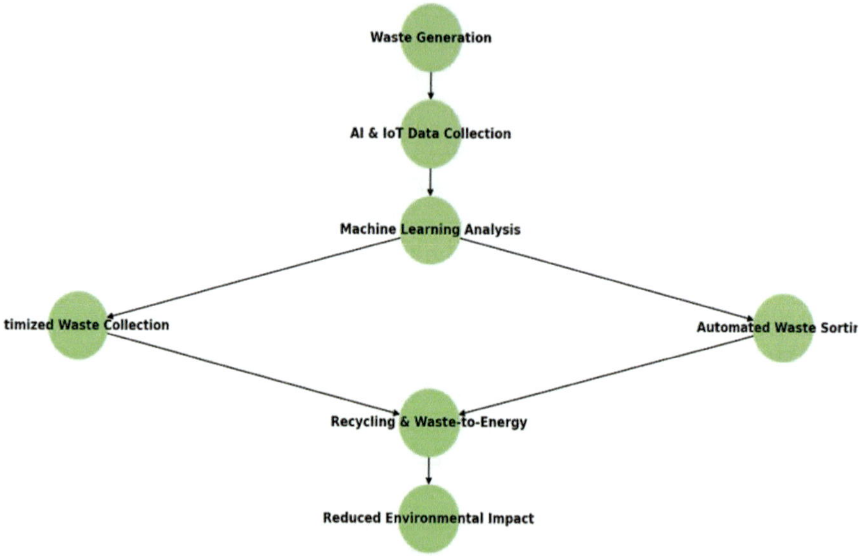

Figure 15.1 AI-based waste management process learning.

15.3.2 Machine for automatic waste sorting

One of the most important components of waste management is accurate waste classification and sorting. Incorrect insulation leads to pollution in recycling systems, which reduces the effectiveness of material recycling processes. Methods of traditional waste sorting include manual labor or basic mechanical separation, both exposed to errors and disabilities. Waste classification accuracy in the ML-driven data view model has improved significantly. Deep learning algorithms, especially fixed nervous networks (CNN), enable the AI system to analyze images of waste and classify materials in different classes such as plastic, metal, paper, glass. and organic waste. These models learn from large datasets with marked waste images, so that they can identify different types of waste with high precision. This technique is integrated into AI-operated robot sorting, which can automatically select and separate waste articles on the conveyor belt in recycled plants. Companies such as Zenrobotics (Finland) and Recyclee (UK) have successfully implemented AI-driven waste sorting of robots, increased recycling efficiency, and reduced pollution rates.

15.3.3 Future production model

The exact forecast for waste production is necessary for the collection of waste, resource allocation, and optimization of landfill management. Predictive analysis models powered by ML use historical data and real-time inputs to predict trends in waste production, allowing municipalities and industries to plan effective strategies for waste management. Time series forecast models such as long short-term memory (LSTM) network, ARIMA (autoregressive integrated moving average), and the Prophet have been used to predict daily, weekly, and seasonal waste production patterns. These models analyze variables such as population density, economic activity, weather conditions, and consumer behavior to estimate future waste versions with high accuracy.

For example, in the city of New York, the ML-based waste spread model made to reduce the landfill overflow by 30% by adjusting collections on the basis of anxious waste production levels. Similarly, smart waste management systems in Barcelona reduced unnecessary waste collection trips by 40%, leading to significant cost savings and lower carbon emissions through the use of sensor-equipped bins and AI-based forecasts.

15.3.4 AI-covered waste collection and timetable adaptation

Effective collection of waste is important to reduce operating costs and reduce the environmental impact of waste transport. Traditional waste collection follows the static plan, which often leads to half-throat or overflow, and discards fuel and labor. The ML-based dynamic root optimization has changed the waste collection by activating real-time decisions based on fill levels, traffic conditions, and weather forecasts.

Strengthening learning (RL) algorithms, such as deep Q-networks (DQNs) and A3C (endurance-profit actor criticism), have been distributed to optimize the waste collection routes dynamically. These models learn the previous collection data and sensor input to recommend the most effective collection roads. For example, Amsterdam's Smart Waste Collection system uses IoT-activated rooms equipped with AI sensors, which inform trucks when they are near capacity. AI then determines the lowest and highest fuel-efficient routes, reduces the waste collection by 25%, and fuel consumption by 20%.

15.3.5 Illegal dumping detection and environmental monitoring

The algorithm has been distributed to detect the AI-operated data view model and deviation to detect illegal dumping places. Satellite images, drones, and AI-enhanced CCTV cameras are used to monitor and identify waste accumulation in unauthorized places. For example, a study that uses U-Net deep learning models demonstrated 98.3% accuracy in identifying illegal dumps from air and satellite images. In South Korea, integrated with AI-based surveillance cameras, rapid object detection models such as R-CNN, have successfully reduced illegal dumping cases within one year of implementation by 35%. Another innovative

A combination of AI-controlled monitoring and blockchain-based waste tracking is also detected to improve transparency in waste management. Blockchain enables officers to detect waste material movement, ensure responsible disposal, and reduce fraud in waste recycling programs.

15.3.6 AI waste-to-energy converting

Waste-to-energy (WTE) technology is an important aspect of permanent waste management, allowing organic and plastic waste to be converted to bioenergy, fuel, or electricity. ML models optimize WTE procedures by improving the efficiency of plastic pyrolysis, anaerobic digestion, and gasification.

For example, the AI models as recurrent nerve networks (RNN) and support vector region (SVR) have been used to predict energy production from different waste compositions, and help operators to meet parameters to maximize efficiency. Studies have shown that AI-enhanced WTE plants receive 15–20% high energy recovery rate compared to traditional methods. AI WTE also helps to monitor real-time emissions, and ensure compliance with environmental regulations on air quality and carbon emissions. By taking advantage of the future model, the authorities can prevent dangerous emissions and optimize the environmental reaction of WTE functions.

Machine learning has transformed waste management into an intelligent, automatic, and durable system of a stable, labor-intensive process. AI-operated waste classification, future indication analysis, route adjustment, illegal dumping detection, and converting waste-to-energy have improved efficiency, cost savings, and environmental impacts greatly. Challenge these develop, challenges such as data availability, regulatory obstacles and implementation costs. Future research should focus on integrated

15.4 Implementation of ML-based Waste Management Systems

15.4.1 ML-based waste management requirements

Data collection system: A successful ML-based waste management system depends on high-quality data. IoT-equipped with competent smart rooms, fill-level sensors, weight sensors, and RFID tags, provide real-time data on waste accumulation. In addition, waste sites are equipped with GPS and traffic sensor collection routes, while CCTV and data vision systems classify waste at recycling facilities. Public commitment data, such as social gripper and insight into social media, increase the efficiency of waste monitoring.. Waste management data is processed using cloud-based servers such as AWS and Google Cloud. For the ML model training, GPU and TPU are required to handle calculations of deep learning.

15.4.2 Software and ML algorithms

Table 15.1 ML-based waste management systems integrate various models

ML algorithm	Application in waste management
Computer vision (CNNs, YOLO, ResNet)	Automated waste sorting
Time-series forecasting (LSTM, ARIMA, Prophet)	Predicting future waste generation
Reinforcement learning (DQN, A3C)	Optimizing waste collection routes
Unsupervised learning (K-Means, DBSCAN)	Identifying illegal dumping sites

15.4.3 ML-based waste sorting and classification

Traditional waste sorting is labor-intensive and subject to errors. ML-operated computer vision models automate the identification and classification of waste, enabling effective separation of plastic, metal, glass, and organic matter. AI-operated robot sorting arms increase accuracy, reduces pollution and operating costs.

15.4.4 Implementation of AI waste sorting

Table 15.2 The process includes three key steps

Stage	**Description**
Data collection	AI-powered cameras scan waste items and process images using labeled ML datasets like TrashNet or TACO
Model training	CNNs differentiate recyclable and nonrecyclable waste, refining sorting accuracy
Real-time classification	AI-powered robotic arms sort waste on conveyor belts in recycling plants

15.4.5 Waste collection and route scheme

15.4.5.1 Challenges in traditional waste collection

This results in unnecessary fuel use, high costs, and poor allocation of resources for waste management. The AI-based root optimization models adjust the dynamic waste plan by using real time without filling levels and traffic conditions

Table 15.3 Implementation of AI route optimization

Stage	Process
Data input	IoT sensors monitor bin fill levels, GPS tracks real-time traffic data
ML model selection	Reinforcement learning (DQN) optimizes routes; genetic algorithms (GAs) improve fleet scheduling
Deployment and execution	AI adjusts collection schedules dynamically; trucks receive optimized routes via mobile apps

Table 15.4 Real-world examples of AI in waste collection

City	AI waste management system	Impact
Amsterdam	AI-optimized waste collection routes	25% reduction in waste collection time
Singapore	AI-driven fleet management	15% reduction in fuel costs

15.4.6 ML-based challenges of waste management and future scope

Despite the benefits, AI-operated waste management faces challenges such as high initial investment costs, data quality restrictions, and environmental regulations. The ML model requires high-quality label data set for accuracy, and privacy regulations for privacy can prevent trace of real-time waste tracking of future scope and innovation: AI-proven wire Waste Tresquity by AI-POILED WASTE Management, FUTURE WAST EFED STALIZ-ITIVARED, WASTEAIAPARED WIFTERED. The purpose of these innovations is to build zero illustrated cities with maximum efficiency in waste treatment. Machine learning changes waste management by automating waste management, adjusting collections, and future analysis. Cities such as Barcelona, Singapore, and Amsterdam have successfully integrated AI-operated waste solutions, reduced operating costs and improved stability. Although there are challenges such as high costs and regulatory barriers, future blockchain progress, IoT integration, and AI-controlled recycling facilities will lead the industry to a zero-up-ended a circular economy. By using ML-based waste management, authorities and organizations can achieve high recycling rates, low carbon footprints, and more efficient waste treatment, which ensure long-term environment and economic stability.

15.5 Real-world ML Application in Waste Management

Machine learning (ML) has traditionally revolutionized waste management by starting automation, data-driven decisions, and intelligent adaptations in the traditionally dependent manual labor and fixed collection plan. World-wide, government and private organizations are using ML-based waste management systems to improve efficiency, reduce costs, and reduce environmental impacts. From automatic waste sorting robots to AI-driven route optimization for garbage collection, ML is transforming how the waste is handled, treated, and recycled. This section examines the study of the real world's case on ML applications in waste management, focusing on successful implementation in different countries and industries. These cases of studies emphasize how AI-driven waste solutions increase the speed of recycling, reduce landfill addiction, and adapt collection processes.

15.5.1 AI-powered waste sorting: ZenRobotics (Finland)

Traditional waste sorting methods depend closely on manual hard work and fundamental mechanical separation strategies. These methods aren't simply inefficient, but also liable to mistakes, leading to infection of recyclable materials and higher landfill dependency. Recognizing the inefficiencies in traditional waste sorting, ZenRobotics, a Finnish AI enterprise, developed an ML-powered robotic waste-sorting machine.

15.5.2 Implementation of machine learning

ZenRobotics applied laptop imaginative and prescient-based totally deep learning fashions to automate the sorting of municipal stable waste (MSW). Their AI-pushed robot arms are equipped with convolutional neural networks (CNNs) educated on heaps of waste pix. The gadget can:

15.5.3 Impact and results

- Sorting accuracy: The AI-powered machine completed a 95% accuracy rate in waste class.
- Increased recycling efficiency: Recycling prices accelerated by 30–40%, decreasing waste despatched to landfills.

- Cost reduction: Manual sorting fees have been reduced by 50%, making the procedure more green and scalable.
- Reinforcement learning algorithms to optimize series schedules dynamically.

15.5.4 Illegal dumping detection: AI-powered surveillance in South Korea

Illegal waste dumping is a major environmental concern in many cities. In South Korea, unauthorized disposal of industrial and hazardous waste led to increased pollution and regulatory challenges. Traditional surveillance methods using human inspections and manual monitoring were ineffective.

15.5.5 Implementation of machine learning

The South Korean government implemented AI-powered surveillance systems with: computer vision models (Faster R-CNN, YOLO) to stumble upon illegal waste dumping from CCTV pictures and drone imagery. Unsupervised anomaly detection algorithms are used to flag suspicious activities. NLP (Natural language processing) models are used to investigate public complaints about illegal dumping.

15.5.5.1 Impact and results

Detection accuracy: The AI system detected illegal dumping with 98.3% accuracy, reducing false alarms.

Faster response times: Authorities had been capable of replying to unlawful dumping incidents 40% quicker.

Environmental benefits: Illegal waste dumping cases decreased 35% within 12 months of implementation.

These case studies demonstrate that ML-driven waste management systems enhance efficiency, lessen charges, and limit environmental impact. From AI-powered sorting robots and predictive analytics to route optimization and waste-to-energy improvements, machine learning is revolutionizing the waste industry. Despite challenges such as high initial costs and regulatory compliance issues, the long-term benefits of AI adoption far outweigh the drawbacks. Future improvements in AI, IoT, and blockchain waste tracking will further power the improvement of zero-waste towns and a circular economy.

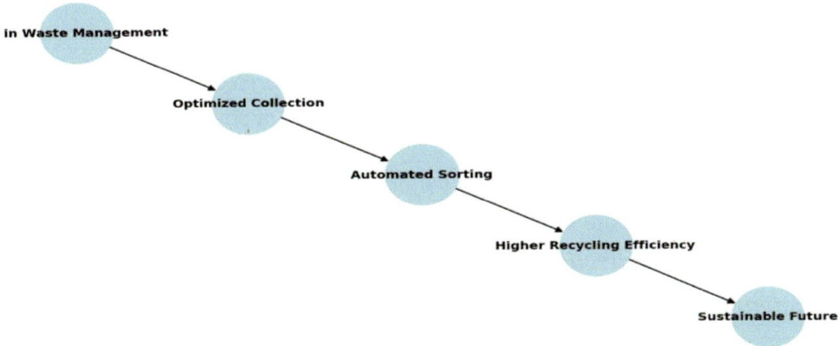

Figure 15.2 AI-based waste management flowchart.

AI in waste management → enables optimized collection (smart routing, automated scheduling).Optimized collection → leads to automated sorting (AI-driven waste classification).Automated sorting → improves higher recycling efficiency (better material recovery). Higher recycling efficiency → contributes to a sustainable future (zero waste, reduced landfills). Figure 15.2 illustrated AI-based waste management flow chart.

15.6 Comparative Analysis: AI and ML in Waste Management – Singapore vs San Francisco

Effective waste management is crucial for urban stability, and Cityworks worldwide is integrating artificial intelligence (AI) and machine learning (ML) to improve treatment, recycling, and collection efficiency. This comparative analysis focuses on Singapore and San Francisco, two cities known for their advanced waste management strategies. The study checks how much each city uses AI and ML-based solutions and evaluates whether their adoption has led to a significant improvement in waste management, recycling speed, and environmental stability.

15.6.1 Singapore: AI-powered smart waste management

Overview: Singapore is a global leader in smart city innovation and has integrated AI into many aspects of its waste collection, recycling, and disposal. With limited land for landfills, Singapore depends on consumers and AI-operated waste management strategies to maintain stability.

15.6.1.1 Use of AI and ML in waste management

Smart requirements and IoT sensor

- The Singapore National Environment Agency (NEA) has equipped the AI-operated smart room with a fill-level sensor that optimizes routes of waste collection.
- The ML-based analysis helped predict the level of filling and dynamic planning, and reduced unnecessary waste collections.

AI-adapted waste collection route

- Strengthening models optimize waste collection, reducing the cost of fuel by 20%.
- Self-driving waste cars are being tested to autonomously navigate waste collection routes, reducing labor requirements.

AI-operated recycling facility

- Machine-based robot sorting systems classify waste with 95% accuracy in recycles and nontransformation.
- AI-operated optical sorts help different plastic types, paper, and metals, improving recycling efficiency by 35%.

Forecast waste analysis

- Time series forecast models (LSTM, Arima) predict the trends for waste production based on seasonal and economic factors.
- ML algorithms analyze waste composition trends to improve waste-to-energy conversion processes.

15.6.2 AI's influence on waste management of Singapore

- The cost of waste collection reduced 30% with optimized AI-based logistics.
- The recycling rate increased from 50% to 65% due to AI-managed sorting and classification.
- The carbon footprint in the form of Ai-Saksham Route adaptation fell by 25%, reducing unnecessary fuel use.

15.6.2.1 Boundaries and future scope

- High early costs: AI implementation requires large investments in infrastructure and smart waste management systems.
- Data privacy concern: IoT-driven waste monitoring raises questions about data security and public transparency.

15.6.3 San Francisco: A stability leader with limited AI adoption

San Francisco, U.S. – One of the most progressive cities, it achieves a landfill submission rate of more than 80%. The city's waste management policy focuses on livestock manure, recycling, and strict regulations on landfill. However, AI adoption in waste management is limited compared to Singapore.

15.6.3.1 Smart waste sorting

- The city's waste management company, Recology, has introduced an AI-controlled sorting system, but on a limited scale.
- Data vision-based AI robots help sort mutual materials, increase efficiency, but do not completely automate the process.

15.6.3.2 Restricted AI-based collection adaptation

- San Francisco still depends on the planned waste collection instead of the real-time AI-operated adaptation.
- There is no implementation of self-driving waste collection cars or AI-based route scheme.

15.6.3.3 Manual recycling and composting system

- More than 50% of waste treatment includes continued manual sorting.
- AI help is available, but not widely used, which causes high labor costs and disabilities.

15.6.3.4 Community-based waste management

- The city instead of relying on AI automation emphasizes waste sorting and civilian participation in livestock manure.
- The ML model is used for waste behavior analysis, but the implementation is not wider as Singapore.

15.6.4 AI's impact on San francisco's waste management

In San Francisco, the landfill derogation rate is high (80%), but AI has not been the most important driver of success. Recycling efficiency increased by 20%, but mostly due to politically driven efforts rather than AI interest automation. Carbon emissions have dropped by 15%, mainly from fertilizer and waste-to-energy initiatives, not AI adaptation.

15.6.4.1 Borders and future scope

AI deficiency in collection adjustment:

Unlike Singapore, San Francisco still follows the manual waste collection schedule.

High addiction to human sorting:

Manual pruning is labor-intensive and expensive, which requires further efficiency for AI-powered automation.

15.6.5 Comparative analysis

Table 15.5 AI-driven vs. traditional smart waste management

Factor	Singapore (AI-driven system)	San Francisco (Policy-based system)
AI usage in waste sorting	High (Robotic AI sorting: 95% accuracy)	Limited (Some AI-assisted sorting, mostly manual)
AI-based route optimization	Fully implemented (ML-driven route optimization, reducing costs by 30%)	Not implemented (Fixed waste collection schedules)
Self-driving waste trucks	Testing phase (Autonomous garbage collection)	Not implemented
Recycling rate increase due to AI	+35% improvement	+20% (Due to policy, not AI)
Carbon footprint reduction	25% lower due to AI optimization	15% lower (Mainly due to composting)
Waste-to-energy efficiency	18% higher due to AI-assisted conversion	Minimal AI impact on energy recovery

15.6.6 The meaning of AI in major conclusions and waste management

15.6.6.1 AI-driven adaptation leads to more efficiency

The integration of Singapore's AI has improved cost efficiency, the degree of recovery, and the environmental ability of waste sorting, collection, and future analysis. On the other hand, San Francisco's policy-fitting approach has been successful with waste reduction, but AI is missing, which can further increase efficiency.

15.6.6.2 Predictive analysis and routing of smart waste with low costs

Singapore's AI-operated waste collection system reduced the cost by 30%, proving that real-time ML-operated adaptation is more efficient than the fixed collection program. If San Francisco adopts AI-based route planning, it can reduce waste collection and fuel consumption.

15.6.6.3 AI-controlled stitching increases the recycling rate

Singapore's AI-driven robotic waste sorting systems achieve 95% accuracy, leading to an increase of 35% in recycling efficiency. San Francisco, still depending on manual sorting, can significantly increase automation to increase the extraction of materials, and reduce labor costs. Singapore's AI-operated waste-to-energy adaptation increased efficiency by 18%, reducing landfill dependence. If San Francisco adopted AI for the waste-to-energy conversion, it could further enhance the sustainability efforts.

15.6.7 Conclusions and future performance

This comparative analysis emphasizes that AI and ML-operated waste management systems provide significant benefits to traditional policy-based approaches. Singapore's AI-integrated system has reduced costs, improved recycling efficiency, and reduced carbon footprint, making it a global leader in smart waste management.

Although San Francisco is very durable, it can benefit from further adoption in collection adjustment, automatic sorting, and improvement of waste-to-energy.

15.7 Future Scope and Innovation in AI for Waste Management

Waste management is an important global issue that affects the environment, public health, and economic stability. The traditional methods for collecting, sorting, and disposal of waste have proven to be inadequate, causing excessive garbage dump, pollution, and resource waste. With the use of artificial intelligence (AI) and machine learning (ML), the industry is experiencing a change, which enables more efficient, data-driven, and durable waste management practices. AI-operated waste management is already making significant advances in areas such as automatic waste sorting, future waste production, adapted waste collection route, and waste-to-energy conversion. This section examines the future scope of AI in waste management, and highlights possible challenges in future innovations, technological advances, and implementation.

15.7.1 Progress in AI-inculcated waste sorting

The AI-controlled sorting system has already shown its ability to improve recycling efficiency. Future innovations will be focused on increasing accuracy, adaptability, and automation in sorting waste. An important development area is a multi-sensor AI sorting system, which uses computer vision, infrared imaging, and robotic weapons to classify waste with near-perfect accuracy. Probably the AI sorting system depends mainly on the trained neural network (CNN) that is trained on the label dataset to identify different types of materials. However, future models will integrate reinforcement learning (RL) and transfer learning, so that the sorting system can be widely allowed to be compatible with new waste materials without the need for wide recovery. This will enable the AI-driven system to handle new materials such as biodegradable plastic and electronic waste components, ensuring that recycling functions remain effective despite changes in physical compositions.

In addition, smart recycling systems will quickly become commonplace, and integrate AI with IoT (Internet of Things) sensors to track the waste composition in real time. These plants will use self-learning AI algorithms that adjust the dynamic sorting mechanism based on real-time response. Such systems will significantly reduce human intervention, increase pruning efficiency, and improve the general quality of recycled materials, making recycling more economical.

15.7.2 AI-competent waste valuation and collection of smart waste

One of the most promising applications of AI in waste management is forecast analysis, which helps municipalities and industries to guess waste production patterns. In the future, the AI models will be even more accurate, invoicing the extra variables such as economic trends, population migration, climate conditions, and consumer behavior to predict waste production. For example, the machine will consider trends on social media and online purchase patterns to predict waste growth during the purchasing season. The municipality can then continuously adjust the waste program to accommodate these ups and downs, prevent garbage dumps, and improve the collection efficiency. The future of the AI-operated waste collection will also include a fully autonomous waste collection vehicle in the future. Self-driving garbage trucks equipped with the AI-enabled navigation system will map the most effective collection routes using real-time traffic data and bin filling level.

These vehicles will communicate with the smart room IoT system, which will automatically inform the collection centers when they are full. This will eliminate unnecessary collection trips, reduce fuel consumption, and reduce the total operating costs.

15.7.3 AI-operated waste-to-energy innovation

Waste-to-energy technology has gained traction as a permanent alternative for landfills, but challenges such as low efficiency and high emissions limited its adoption. In the future, AI will play an important role in adapting the WTE conversion processes, and ensuring that more waste will convert to clean energy with minimal environmental effects. The AI model will be used to adjust the temperature, pressure, and chemical compositions accurately to dynamically adjust plastic pyrolysis, anaerobic digestion, and gasifying techniques. These adjustments will maximize energy production by reducing carbon emissions. In addition, the recurrent nerve network (RNN) and supporting vocal machines (SVMs) will be implemented to predict energy production efficiency based on the upheaval structure coming up, enabling the real-time optimization of the waste-to-energy process.

Another groundbreaking innovation will be AI-driven carbon catch integration in WTE plants. The AI system will analyze future AI system emissions and automatically regulate the exhaust filter to capture and store CO_2 more efficiently. This will help waste-to-energy plants to comply with strict carbon neutrality rules, while making the process more environmentally friendly.

15.7.4 AI for managing e-waste and a circular economy

Electronic waste (e-waste) is an important environmental challenge due to its complex structure, including toxic metals, rare earth elements, and non-biodegradable plastic. The current e-waste recycling methods are inefficient, causing the loss of valuable materials and increasing pollution. AI is ready to bring revolution in e-waste management by activating automatic e-waste washing and material recovery. These systems will benefit from computer vision and X-ray mechanism to identify internal components, allowing accurate extraction of gold, silver, copper, and lithium. This approach will reduce the environmental impact of e-waste when making electronic recycling more profitable. These platforms will use AI-controlled recommendation systems to match businesses that generate garbage with companies that can reproduce these materials. For example, the AI model production companies can connect concrete waste with manufacturers that can process it for new building materials and reduce landfill addiction.

15.7.5 Blockchain and AI integration for waste tracking

One of the biggest challenges in waste management is the lack of transparency and responsibility in waste management. Illegal dumping, poor waste tracking, and disabilities in recycling supply chains often cause environmental damage. A convergence of AI and blockchain techniques will appear in the future to create tamper-proof tracking systems. Blockchain technology will be used to register each step in the waste management process from waste production to final disposal. AI will analyze this data in real time, discover deviations in waste patterns, and mark potential illegal dumping activities. For example, the AI algorithm can detect suspected nonconformities in industrial waste reports, and automatically notify the environmental authorities of potential violations of regulations. In addition, smart waste contracts run by blockchain and AI will ensure that companies follow permanent waste management practices. These smart contracts will automatically implement recycling obligations and free up penalties for nontransport. This integration will promote a more transparent and responsible waste management ecosystem.

15.7.6 AI and the robotics sea in cleaning waste

Marine pollution is an increasing crisis, where millions of tons of plastic waste are accumulated in the seas each year. Future AI innovation will focus on autonomous oatmeal robots, which are able to detect, collect, and treat liquid waste. These robots will be equipped with AI-operated object recognition systems that can distinguish between marine life and plastic residue, ensuring that the clean-up effort does not damage the ecosystem. This will enable cleaning operations to target high-risk areas and maximize efficiency. In addition, the AI-powered microplastic detection system will be distributed to track and filter microplastics particles from water, which will address one of the most pressing environmental challenges of our time.

15.8 Conclusion

The implementation of ML-based waste management systems marks a significant shift toward more data-powered, automated, and sustainable approaches, compared to traditional waste management methods.

AI-operated solutions, such as robotic waste sorting, waste analysis, and smart waste collection optimization, have already shown their efficiency in reducing landfill dependence, reducing carbon emissions, and increasing recycling velocity. Comparative studies between Singapore and San Francisco can increase Singapore's evergreen waste-to-energy conversion. Waste management includes the main benefits of AI, but there are challenges, including high infrastructure costs, regulatory barriers, and advanced data collection methods. Using blockchain, autonomous waste collection vehicles and AI-operated cleaning of ocean waste will provide further revolution in the future progression of AI-operated waste. In fact, a permanent waste management system, authorities, industries, and researchers should work together to integrate AI-powered solutions into national waste policy, working on technological innovations. AI-operated waste management, by embracing waste managementsolutions, can help cities around the world achieve a zero-waste future, promoting both stability and efficiency.

References

World Bank (2018). *What a Waste 2.0: A Global Snapshot of Solid Waste Management to 2050.*

ZenRobotics (2022).*AI-Powered Waste Sorting in Smart Cities.*

National Environment Agency, Singapore (2023). *Smart Waste Collection Using AI and IoT Sensors.*

Gupta, R., et al. (2021).*Deep Learning for Waste Sorting: A Convolutional Neural Network Approach.* IEEE Transactions on Sustainable Computing.

Thung & Yang (2020). *TrashNet: An Open-Source Image Dataset for Waste Classification Using AI.*

Ramalingam, S., et al. (2022). *A Comparative Study of AI-Driven Waste Sorting Algorithms.* Journal of Environmental AI Research.

Kourtis, P., et al. (2019). *Predictive Analytics for Municipal Waste Generation Using LSTM Models.* Waste Management Journal.

Shah, A., et al. (2021). *IoT-Based Smart Bin Sensors and AI Prediction Models for Waste Management.* Smart Cities Journal.

Mehta, D., et al. (2023). *Using Social Media Sentiment Analysis for Waste Overflow Prediction.* AI for Sustainable Development.

Lee, J., et al. (2021). *AI-Based Fleet Management for Smart Waste Collection.* Transportation and Logistics in Smart Cities.

Olivier, M., et al. (2021). *AI-Powered Drone Surveillance for Illegal Waste Dumping Detection.* Journal of Environmental AI.

Santos, G., et al. (2023). *Satellite Imaging and AI for Ocean Plastic Waste Monitoring.* Marine AI Research.

Ding, Y., et al. (2020). *Reinforcement Learning for AI-Powered Waste Collection Optimization.* International Journal of AI in Smart Cities.

Garcia, H., et al. (2022). *Genetic Algorithm-Based AI Models for Traffic-Aware Waste Collection Routes.* Sustainable Urban Planning Journal.

United Nations Environment Programme (2023). *Future of AI in Waste-to-Energy and a Circular Economy Approaches.*

Index

About the Editors

Rakesh Kumar is an eminent educationist and decorated army officer with 26 years of experience. His areas of expertise are finance, business law, labor law, technology, and his teaching concentration includes finance, business law, labor law, technology. He is an interdisciplinary researcher. He is also involved in the course and syllabus design and development of various management topics, including an analytics and management curriculum. He served in the Indian Army for more than 23 year in management, academic, operational, and training. He has published more than 75 research papers in leading databases. He is the editor of four books. He has published 33 Indian Patents. He has been part of the board of study, board of examination, and the evaluation department in government organizations. He has completed M.COM, MBA (2 years PGDM), PGDCA, PGDBM, etc. He is also pursuing an MA (Economics) and is a company secretary. He was part of Uttaranchal University and is currently at the Research faculty in Graphic Era Deemed to be University.

Sachi Nandan Mohanty was recognized as a top 2% world scientist by Stanford University and Elsevier for the year 2023. He received his PostDoc from Indian Institute of Technology Kanpur, India in the year 2019 and his Ph.D. from Indian Institute of Technology Kharagpur, India in the year 2015, with an MHRD scholarship from the Government of India. He has authored/edited 42 books, published by IEEE-Wiley, Springer, Wiley, CRC Press, NOVA and DeGruyter. He was awarded the best thesis award first prize by the Computer Society of India in the year 2015. He has guided 9 Ph.D. Scholars and 23 postgraduate students. He has published 241 papers in international journals of repute. He has been awarded many awards and fellowships during his career including Fellow of Institute of Engineers, IETE, Ambassador European Alliance Innovation (EAI), and Senior member of IEEE Computer Society Hyderabad chapter.

Isaías Scalabrin Bianchi holds a Ph.D. in Technology and Information Systems from the University of Minho, Portugal, and has served as a visiting researcher at the University of Twente in the Netherlands. He is a civil servant at the Federal University of Santa Catarina, where he has held various management roles since 2010. Additionally, he serves as a professor in the Master's Program in Management at the same university. His research interests include IT governance, information technology business process management, business and artificial intelligence, and digital transformation and innovation, with a particular focus on public organizations and higher education institutions. His research has been published in prestigious IS conferences and international journals. Since 2020, he has also worked as a visiting professor at Al-Farabi Kazakh National University, where he teaches and supervises Ph.D. students in the areas of digital transformation, artificial intelligence, digital marketing, and supply chain management. He leads research projects funded by the Ministry of Education of Kazakhstan. Additionally, he co-supervises master's students at the University of Minho and Instituto Universitário de Lisboa. Isaías has achieved Moodle Educator Competence through the Moodle Educator Certification Program. Furthermore, He has edited and published three books with Springer and has also served as a guest editor for several indexed journals

For Product Safety Concerns and Information please contact our
EU representative GPSR@taylorandfrancis.com Taylor & Francis
Verlag GmbH, Kaufingerstraße 24, 80331 München, Germany